OZARK FOREST FORENSICS

OZARK FOREST
FORENSICS

*The Science Behind the Scenery
in Our Regional Forests*

FREDERICK PAILLET AND
STEVEN STEPHENSON

The OZARK
SOCIETY
FOUNDATION

2019

Designed by Liz Lester

⊗ The paper used in this publication meets the minimum requirements
of the American National Standard for Permanence of Paper for Printed
Library Materials Z39.48-1984.

Publisher: Ozark Society Foundation, PO Box 2914, Little Rock, AR 72203
The Ozark Society Foundation publishes books and guides on nature and the
environment of the Ozark-Ouachita mountain region. Purchasing information
available at www.ozarksociety.net.

CONTENTS

PREFACE

This book originated from our participation on outings with the Ozark Society, local chapters of the Sierra Club, and other outdoor organizations. We are both university professors and natural scientists in the geosciences or biology. As such, we are regularly asked questions about the things we see on outings in the Ozarks. Moreover, one of us (Paillet) is an inveterate sketch artist who finds scenes of specific technical interest for illustration while on the trail, which naturally leads to questions about why any given subject is worthy of documentation. Introducing these scientific insights to eager and enthusiastic outdoor audiences has shown us how much general interest there is in understanding the landscape around us. Here is our attempt to satisfy that interest.

Our presentation of forest forensics in the Ozarks addresses these natural history insights at the large-scale landscape level and from the perspective of a typical Ozark woodland outing. All of our individual lessons are illustrated by sometimes simplified line drawings, specifically designed to be unambiguous in their interpretation by the non-expert, without the visual clutter typical of photographs made in the field. Most of our illustrations and diagrams are based on field sketches or photographs from locations in the Ozarks of Missouri or Arkansas, with specific locations mentioned only when relevant to the discussion. We avoid using obscure technical jargon as much as possible, but some special terms are necessary. We have tried to provide just enough terminology that any reader interested in more background on a specific subject can easily use our descriptive terms to find such information on the Internet.

Although ostensibly addressing the forests of the Ozarks, our book covers far more than just forestry and trees. We deal with the geology, climate, and biology of an integrated ecosystem. This book is sponsored by the Ozark Society Foundation, and we address some conservation issues in our brief concluding chapter, but only insofar as the science we have presented influences the wider arena of conservation, with all its complications in a world full of competing and conflicting interests. We have also taken the liberty of including an extra chapter devoted to species of special interest to us and to many other local outdoor enthusiasts. These are species that loom large in Ozark folklore (chinquapin and ginseng), are notable for their extreme rarity (ladyslipper orchids), or are just plain interesting (paw-paw and farkleberry). We hope that our foray into "forest forensics" can add to the pleasure of a simple walk—to the many walks that await us in the Ozark forests at our doorsteps.

ACKNOWLEDGMENTS

The manuscript benefited from the helpful review comments of Don Culwell, Helen Davis, Sara Fitzsimmons, Graham Hawks, Judi Nail, Janet Parsch, Tom Perry, and John Van Brahana. Their help in catching oversimplifications, identifying inadequate explanations, and just tracking down garden-variety typographical errors significantly contributed to the quality of the finished product.

CHAPTER 1

A Walk in the Woods

Introduction to Forest Forensics

A typical walk in the Ozark woods often starts at a roadside trailhead with a well-marked path leading into the forest. Some of the most pleasant days in the woods are in autumn, with bracing temperatures and increasing visibility as leaves fall from branches. The stimulating exercise, pleasant air, and natural surroundings may initially motivate us to get out on the woodland trail. But sooner or later, most of us want to know a little more about the things that go on in the forest we are passing through. For example, you might become aware of the many interesting bark patterns—fissures, plates, or peeling strips— that characterize the largest trees along the trail. In fact, if you look closely, it is not just the bark itself that is the attraction: it is also the diverse community of mosses and lichens that grows on top of the bark, and the callus tissue that develops as trees accumulate scars from their battles with insects, wildlife, and the elements. Perhaps you notice changes in the texture of the tree growth as you move from an open forest of tall, columnar trees into an area with numerous smaller trees that seem to be of a different variety—and that are perhaps entangled in grape vines or, somewhat less to your hiking pleasure, greenbrier and poison ivy. You wonder what could account for the difference and speculate on how it might have come about. Taking a close look at the ground underfoot, with its thick layer of leaf litter and woody debris, you might notice different depths of leaf accumulation associated with certain locations, and a near absence of litter at other locations where there is a plush layer of mosses or other types of ground cover. You might even notice little furrows carved in the leaf litter—the result of armadillos plowing the ground for grubs and other delicacies. The more you look, the more you can find to wonder about. This little book is written to guide you in a more systematic investigation of the Ozark woodlands. Our intent is to make an ordinary walk in the forest all the more interesting and stimulating as you learn to read the many stories of past events and the fiercely competitive struggles that are occurring all around you.

The discipline of forest forensics had its beginnings in New England, and its founding father was none other than Henry David Thoreau, as reflected in his *Journal* and in the chronicles of his forays in the Maine woods and on Cape Cod. Thoreau famously noted the extent to which human activities had transformed the landscape around him, lamenting that the largest wild animal left in the environs of his native village of Concord was the muskrat. Nonetheless, he undertook a systematic study of the natural world, noting the locations where certain species of trees were growing and speculating about changes that might occur in the forest as trees aged and were replaced by others that seeded in beneath them. He laboriously mapped the bathymetry of Walden Pond and discovered, to his surprise, that when you count tree rings in a felled forest giant, the size of the tree may not be related to its age at all. Many of Thoreau's observations seem a way of entertaining himself on his walks around rural Massachusetts. With a little tutoring, you, our reader, can learn to entertain yourself in the same way—by paying careful attention to the natural world while enjoying the outdoor resources we are blessed with in the Ozarks.

Thoreau's collection of observations is full of detail but mostly lacks any centralized organization. The science of forest forensics was subsequently begun in a more structured way by Robert Marshall. A personal hero of one of the authors, Marshall is known as the father of the wilderness concept. Following in the footsteps of John Muir and Theodore Roosevelt, he recognized the intrinsic value of preserving wilderness for its recreational enjoyment and emotional stimulation. The Bob Marshall Wilderness Area in western Montana, affectionately known by passionate backpackers as "The Big Bob," is a suitable tribute to this wilderness advocate. Located along the Continental Divide, amid fabulous mountain scenery, it contains the largest free-flowing river in the lower forty-eight states. Marshall was one of the principal founders of the Wilderness Society, and his name will forever be associated with the concept of wilderness for its own sake. But long before he achieved that fame, he practically invented quantitative forest forensics. In 1924, working toward his master's thesis at Harvard Forest in central Massachusetts, he laboriously reconstructed the life histories of trees in a hemlock-dominated forest, speculating about the previous history of the forest and the past events that influenced the individual trees he was investigating.

Bob Marshall's ambitious career plan did not include further studies of this type, and his early work remained unpublished for many years while he went on to achieve fame in other ways. However, the concept of such reconstruction remained alive at the Harvard Forest, to be fulfilled in a monumental study by Earl Stephens in the 1950s (although this insightful data set, too, remained unpublished, until another researcher took the time to cast its wealth of historical information into a coherent story).[1] Stephens's research involved the intense reconstruction of every detail of tree growth on a forested area of about a third of an acre. The results would set the standard for many such investigations, though the approach has expanded as techniques of forensic science have progressed to include quantitative tree-ring analysis, isotopic studies, pollen analysis, statistical forest-growth models, and radiometric age dating.

Now let's consider a typical walk in an Ozark forest to see what insights can be achieved through techniques pioneered by Marshall and Stephens. A typical trail starts out on the sandstone plateau formed by the sedimentary layers that provide the foundation for the Ozark landscape (Figure 1.1). The hiker steps into an open woodland composed of trees with roughly corrugated bark, and many of these appear to be some variety of oak. But exactly what oaks are they,

---

1. For those interested in the history of forest forensics, we recommend this landmark study: C. D. Oliver and E. P. Stephens, "Reconstruction of a mixed-species forest in central New England," *Ecology* 58 (1977): 562–572. Some of the most recent reconstructions of forest history are in publications that you can download from the Harvard Forest web page (http://harvardforest.fas.harvard.edu/research-publications). Other important early studies in forest forensics and the reconstruction of deciduous forest history in North America can be obtained by consulting the publications of Professor Craig Lorimer of the University of Wisconsin.

and are there specific varieties that grow in this kind of habitat? You notice that the trees are not especially tall in this location, and so a good deal of light penetrates to the forest floor. As a result, there is a nearly continuous distribution of low-growing woody shrubs, adapted to this specific environment and representing their own biological community. The carpet-like distribution of these plants suggests that many did not originate from seeds but rather were propagated through lateral expansion of the root system (and there is much to learn from the dichotomy between reproduction by seed and by vegetative propagation via underground runners or root sprouting). Then, if you look closely at the trees, you can see broken branches and other damage to many of them, along with scars and sections of missing bark along their trunks. Each of these tells the story of some event in the past, and an inventory of such observations can provide a local history of windstorms, fires, and diseases. We could take tree-ring cores to determine the trees' ages—did they originate in a single event or when older trees were toppled in individual windstorms at many different times? We could also look for abrupt changes in growth rate that were produced by disturbances that occurred many years ago. Often, this kind of elaborate study is not even needed because a tree's shape, branching pattern, and scars give a rather good idea of the history it has experienced. Thus, there are many things we can learn by simply noting the size, shape, patterns of damage, and species composition of the forest in this upland location.

One particular focus of study in upland forests is the nature of the leaves that serve as the fundamental energy source—the photo-synthesizing solar collectors—for the forest ecosystem. Leaf size and shape are often taken as a diagnostic cue in determining the identity of a species, but leaf ecology involves a whole lot more than that. For example, leaves need to be defended against all kinds of predators. They also have to be oriented to maximize energy efficiency, and that orientation depends on the location of leaves with respect to canopy structure—which, in turn, involves tree architecture as influenced by branching pattern and inherent growth form. This is largely a matter of resource allocation. Should the tree try to stay above the competition by outgrowing the surrounding trees, or should it let competitors waste their energy on building wood and instead concentrate on

FIGURE 1.1. Open oak–hickory forest on a level upland sandstone plateau, showing typical bark textures, damage to trees, fallen branches, the low shrub layer, smaller understory trees, and fallen logs—all of which are indications of processes at work in a maturing forest.

making its own leaves more efficient in partial shade? Leaves have to find a compromise between opening pores to facilitate intake of carbon dioxide and minimizing loss of water through evaporation. This compromise involves issues of leaf size and shape. This topic turns out to be so involved that a whole scientific discipline has been based on interpreting past climates strictly on the basis of fossil leaves, yielding estimates of atmospheric carbon dioxide, relative humidity, and mean annual temperature. You should be aware, as you walk down the trail, that the pleasantly dappled light filtering through the trees overhead results from an intricate and complicated set of physiological calibrations and is not just a delightful byproduct of the scenic forest environment.

Most Ozark trails will eventually descend through rock ledges formed by the hard sandstone and limestone framework of the landscape (Figure 1.2). The shallow soil, bare rock surfaces, exposure to wind gusts, excessive drainage, and unstable talus on these steep slopes provide a unique environment with its own assemblage of species. Many of the trees that grow here are adapted to this environment, while other, more common trees adopt a contorted and stunted growth form to cope with the stressors they encounter. One byproduct of these conditions is slow growth, such that relatively unimpressive trees on ledges can be much older than large trees in the surrounding forest. The harsh environment also makes these trees very sensitive to climate swings, so they are repositories of rainfall and frost data for the tree-ring analyst. You will note that not one rock-ledge habitat suits all. The local topography forms what ecologists call "microhabitats." There will be hollows in the rock that collect leaf litter, and adjacent outcrops where trees and shrubs are forced to gain a foothold in crevices. During the spring, you can often trace a line of flowering shrubs (white for serviceberry and pink for azalea) that follows the crest of these rock ledges where they wend their way across the opposing hillside. The shallow soils promote a microhabitat that has a lot in common with the harsh desert environment, so it is no surprise that we encounter species, such as prickly

FIGURE 1.2. Trail passing down through rock ledges into the sheltered habitats below, where other tree species join the oaks and where rock slabs are inhabited by lichens, mosses, and various herbs.

pear cactus and gayfeather, that are familiar plants in the arid steppes of Colorado and New Mexico. The dry conditions also greatly retard the decomposition of wood, such that relicts of past forests can persist for centuries. Ecologists even get a bit of help from industrious pack rats that have a habit of collecting a representative sample of the surrounding vegetation in the midden deposits they assemble in rock shelters beneath ledges. This is not just a local Ozark phenomenon. A whole range of critters, including Eurasian hyraxes and South American rodents, have evolved this midden-construction habit—generating, in turn, a cadre of midden-interpretation specialists in research institutes around the world.

When the hiking trail eventually descends through rock outcrops, the hiker encounters yet another environment, often filled with an entirely different set of trees and other plants. In contrast to the exposed rocks above on the hillside, this environment presents a rich habitat where soils accumulate to form substantial thicknesses of organic-rich deposits, and where water filtering down from above maintains soil moisture supplies throughout the year. Just as on the rock ledges above, this location is characterized by a great diversity of microhabitats, including muddy seepage areas adjacent to intermittent springs, giant moss- and lichen-encrusted sandstone slabs, and exposed soil on slopes undercut by steeply tumbling water courses (Figure 1.3). Even the untrained eye recognizes that many of the trees here differ from those on the plateau above, in both growth form and species composition. Some are canted at odd angles and have exposed roots, indicating where they have been affected by slope instability. The shrubs and the herbaceous growth underfoot have their own special character. This deeply shaded environment provides a number of stark contrasts in scenery, representing the interaction of two dramatically opposing trends: moist soils and protected environments encourage large, long-lived trees adapted for life in a comfortable location; while other, more opportunistic trees are adapted to take advantage of the massive disturbance events when steep slopes collapse and periodic floods scour headwater stream habitats. You even find that a distinct odor pervades this habitat as a result of characteristic species such as spicebush and fragrant sumac.

Trails descending through ravines and hollows eventually spill

FIGURE 1.3. Lower slope and headwater drainage with sandstone blocks, exposed ledge in a streambed, an abundance of large shrubs such as spicebush, and small seepage springs.

out onto the alluvial flats along such mountain streams as the upper Buffalo River, Lee Creek, or Frog Bayou (Figure 1.4). Once again the landscape indicates a great transformation, through both the difference in the nature of the plant community and the abundant evidence of human activity in what was the most fertile real estate available to early settlers. One obvious observation is the patchiness of the stream-side forest, which results from a combination of factors: piecemeal abandonment of fields and pastures, periodic migration of the stream channel, and inherent variation in substrate ranging from former sandbar to gravel benches and rich alluvial soil. Stone walls, rusted wire from old fences, and the occasional abandoned well indicate the locations of long-forgotten homesteads.

Another legacy of human occupation is the presence of alien plant species that were introduced deliberately as ornamentals or inadvertently through propagation in livestock fodder. Perhaps the most noticeable aspect of the landscape is the great variety of flood-related damage to trees, including exposed root tangles where trees are

FIGURE 1.4. Stream setting with point-bar and cut-bank habitats, showing evidence of intense scour and prominent flood damage to streamside trees.

undercut by eroding banks, great patches of bark abraded from the bases of sycamores and sweetgums, and multiple-stemmed clumps of trees arising from the broken stump of a former streamside giant. You will probably notice substantial areas of seemingly uniform forest, consisting of evenly spaced and evenly sized specimens of a single species such as sweetgum or black walnut that obviously originated at the same time, as a result of conditions that prevailed many years ago. All these observations are clues to how the past is reflected in the present and how events both man-made and natural echo down through the years.

After ascending from the slopes and alluvial deposits, the typical hike ends as it began—on the edge of a sandstone plateau, back at the original trailhead. One last panoramic view over distant slopes and ridges gives a snapshot of the Ozark landscape (Figure 1.5). This landscape is the result of about half a billion years of deposition and erosion of sediments deposited on the low-lying edge of one of those crystalline rock islands known as "cratons" drifting about in a sea of slowly moving and subducting basalt, the process we know as "continental drift." These mostly flat-lying layers reflect the pulses of mountain building and erosion going on just to the south, as well as the regular fluctuations in sea level that pervade geologic time. The rugged Ozark landscape we see today has resulted from uplift and erosion of those layers, with uplands and prominent ledges formed by the most erosion-resistant layers in the sedimentary sequence. Erosion has effectively etched the topography by selectively removing more erodible layers and leaving the harder rock layers to protrude as ledges and bluffs. But we can see that sandstone or limestone ledges are far from continuous, just as the nature of deposition at any one time in the past was never uniform throughout the landscape. And thus we see evidence of how local habitat conditions in long-forgotten ecosystems can still influence events in the landscape today. Nature seems to have celebrated complexity and diversity through all time scales in the immense span of the history of life on our planet, as if engaged in some grand conspiracy designed to instill wonder in the naturalist.

FIGURE 1.5. View across distant ridges: a generalized image of the Ozark mountain landscape as shaped by millions of years of episodic deposits of harder and softer layers, uplift, and erosion.

This little volume will give the ordinary hiker enough lore about natural events in the Ozark landscape to form an immense appreciation of the observations that can be made during a typical walk. We start with a chapter on the geology that shaped the Ozark Mountains at both the broad regional scale and the individual outcrop scale (Chapter 2). We then introduce the characteristics of the general forest cover of our area, which is nominally classified as the "oak–hickory" forest association (Chapter 3). This is followed by an exploration of the other enclaves of forest scattered within the oak–hickory domain that are dominated by other species, with a relatively minor oak presence (Chapter 4). This brings us to a discussion of forest disturbance history and the way in which presettlement disturbance history might have influenced what we reconstruct as the original prehistoric forests of the Ozarks (Chapter 5). We then discuss how to recognize the ways in which trees tell their stories, both by observing damage history and crown structure and by looking at tree-ring series—from which a lot can be determined by simple observations of the cut stumps and logs regularly encountered along maintained trails (Chapters 6 and 7). The next chapter deals with the flow of water by describing streams, vernal pools, river channel geometry, and karst geology, with emphasis on how these substrates influence the local ecosystem (Chapter 8). Subsequent chapters deal with shrubs and vines, interesting but uncommon species, and the cycle of wildflowers through the year (Chapters 9, 10, and 11). Two chapters present the important topics of fungi, soil microorganisms, and diseases and pathogens in the forest (Chapters 12 and 13). The final chapter discusses alien invasive species and other conservation issues that have important implications for the dynamic landscape we are here to appreciate (Chapter 14). We hope this introduction to the complex world of forest ecology and geomorphology can transform a simple walk down Ozark mountain trails into a real adventure in the exploration of our amazingly complex natural world.

CHAPTER 2

The Geological Foundation of Ozark Forests

Ozark forests are situated on the Ozark uplift, a dome-like swelling of geologic strata that forms the underlying foundation for the rugged scenery we enjoy in our region. Many guidebooks introduce this geology by showing one of several different standard sequences of named formations, displayed in a neat, layer-cake format. In our experience, this simplistic representation is extremely frustrating for the average reader because these formations are often difficult for the nonspecialist to recognize in the field. The fact is that the geology on the ground is far more complicated than the simple way it is represented in such diagrams. This exquisite complexity should be embraced as one of the greatest wonders of the natural world. Ordinary energy-balance arguments would suggest that a rocky sphere orbiting a constant energy source like the sun should have a very boring story to tell. Just the opposite is the case, in a geologically miraculous way. Earth's climate has varied between frozen and desiccated extremes, with the land that would become the Ozarks rising above and sinking below sea level numerous times in between—generating the intricate geological variation observed today. The layers that we see exposed in our landscape represent about three hundred million years of Earth history, a span of time beginning when larger multicellular life first arose and ending just before dinosaurs ruled the land. A deep appreciation of the Ozark landscape should be based on a celebration of the wonderfully diverse series of events this complex history records.

Our understanding of Earth history in the Ozarks and elsewhere is based on the recognition of plate tectonics, whereby continent-sized rafts of relatively low-density aluminum- and silica-rich rocks are buffeted around by the excruciatingly slow overturning of heavier, iron-rich rocks that form the bulk of Earth's outer mantle layer. Careful geological research has shown that this process alternately pushed the continental rafts into a single supercontinent and then caused that supercontinent to fragment and break apart. During much of the period when the rocks we see in the Ozarks were deposited, our region was located on a supercontinent known as Pangaea, which included the landmasses of North America, South America, Africa, and Europe (Figure 2.1). The environment was far different from that of the Ozarks today—in terms of climate, geographic location, relatively low landscape elevation, and the nearby presence of ocean inlets. The most comparable present-day habitat may be that of the Persian Gulf area, but there are some important differences to keep in mind. First, we should note that the modern landscape one can see today in the Persian Gulf region has been deeply affected by the hand of man. For example, when one of the authors worked in the Kuwait desert, there was hardly any vegetation at all beyond a few annual weeds. But a two-square-kilometer research area maintained with complete livestock exclusion for many years had about 50 percent vegetation coverage and resembled the landscape we can see today in many Nevada desert basins. Then we have to remember that the Earth's climate changes rapidly on the geologic time scale, such that there have been many rhythmic cycles of climate. Even the premodern Persian Gulf landscape would have varied significantly during the time that aboriginal humans were living there. Readers should keep in mind that the analogy of the ancient Ozarks with the Persian Gulf region is appropriate but should not be taken too literally.

In the ancient Ozarks, sediments were laid down in various depths of water and then periodically uplifted and partly eroded before another cycle of deposition. While sea level changed, the region was also being uplifted by movement on deep-seated faults in the rocks below. Sometime after three hundred million years ago, the land that would become California and much of Nevada began to be added to the west coast of North America. This caused the future Ozark region to remain above sea level while this tectonic activity uplifted the entire midsection of the continent, and subsequent erosion etched out the rugged terrain we recognize in the topography of our region (Figure 2.2). Thus, the Ozark Mountains are actually an

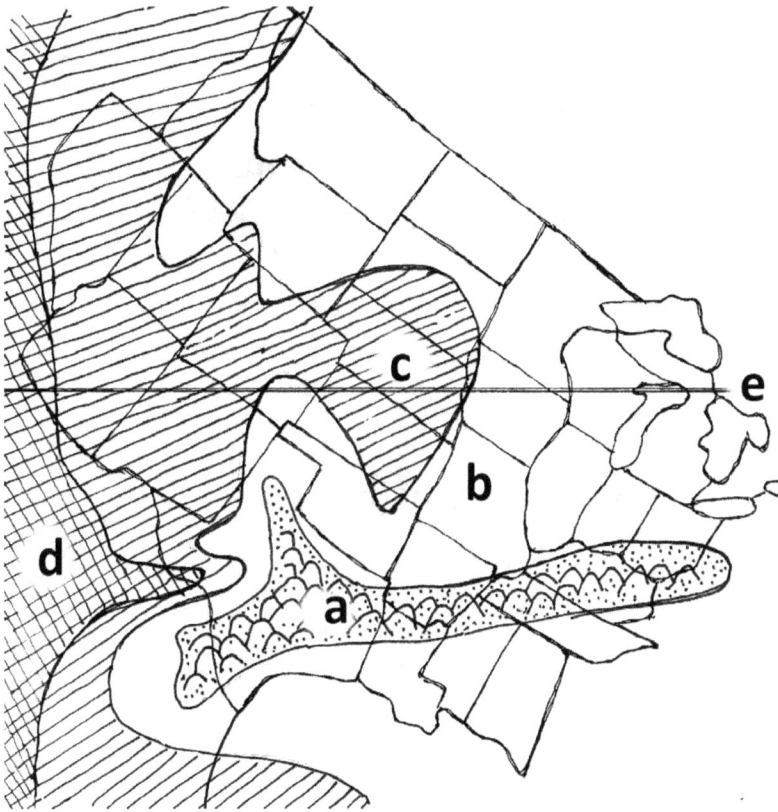

FIGURE 2.1. Ozark regional topography about three hundred million years ago, when the Ouachita–Appalachian mountain chain (**a**) was delivering sediment to a depositional basin on the north flank of the uplift (**b**), with shallow ocean over continental crust to the west (**c**) and deep water over ocean crust where California lies today (**d**). At that time the equator passed diagonally across the North American continent (**e**).

uplifted plateau rather than mountains—more like the canyonlands of the American Southwest than the Blue Ridge Mountains of Virginia. Three intervals within that dome of earth provided especially hard layers for erosion, creating four regional plateaus when the underlying granitic "basement rock" exposure is included. The Boston Mountains Plateau is the youngest of these plateau surfaces and is composed of sandstone in north-central Arkansas. The Springfield Plateau extends over the northwest corner of Arkansas and the southwest corner of

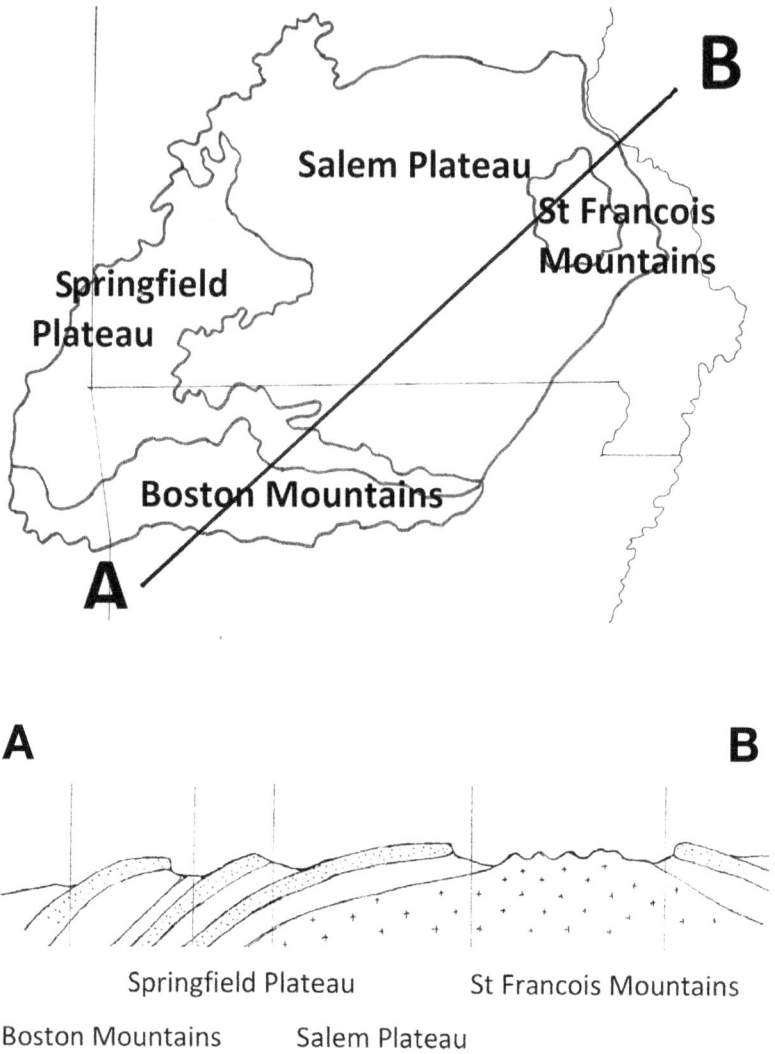

FIGURE 2.2. Map of the Ozark Plateau (**A**) indicating the structure of the Ozark dome and the respective locations of the Boston Mountains, the Springfield and Salem plateaus, and the St. Francois Mountains in its core; and schematic cross section (**B**) showing how the level of the Ozark Plateau is controlled by the presence of relatively erosion-resistant layers (stippled) in the sedimentary strata draped over the crystalline bedrock dome of the Ozark uplift.

Missouri. Even older dolomite forms the Salem Plateau in south-central Missouri, and the St. Francois Mountains represent the center of the uplift where the underlying basement bedrock is exposed.

There is a single fundamental question about Ozark geology that is so obvious we routinely overlook it, just as we take the existence of the Mississippi River for granted. The fact is that the pairing of the Ozark Plateau and Ouachita Mountains in Arkansas is part and parcel of the Appalachian system that pairs the Allegheny Plateau and Appalachian Mountains in Pennsylvania and the Cumberland Plateau and Blue Ridge Mountains in Virginia and Tennessee. So it seems strange that a major river would decide to exit a continent through a mountain chain, isolating the Ozarks from the rest of the system. After all, the Amazon drainage began to flow east only after the Andes Mountains on the western rim of the continent blocked outflow in that direction. Arkansans were given early notice that something serious was going on underneath the Mississippi Valley by the New Madrid earthquake series of 1811–1812. This remains a real mystery today despite all our insight into plate tectonics and our ability to probe the deep Earth with seismic-wave methods. Subsequent studies of depositional horizons in buried Mississippi sediments show that such major earthquakes have been occurring at about five-hundred-year intervals. Deep seismic surveys show that there is a complex, down-dropped, trench-like feature embedded in the top of the deeper crystalline rocks—what geologists call the "basement." This is interpreted as an ancient rift that once threatened to separate the southeastern part of the continent from the rest of North America. Some kind of later activity on this deeply buried feature could have caused the place where the rift intersected the southwestern section of the Appalachian chain to drop down enough to create the Mississippi River pathway. Or the deeply seated faults could have allowed volcanic activity and rising magma (think of diamond emplacement)[1] to swell this intersection with heat, such that it

---

1. For those readers not familiar with Arkansas state parks, the only diamond-bearing volcanic deposit (a volcanic intrusive formation known as a "kimberlite") in the lower forty-eight states is located in southwestern Arkansas and has been developed into the Crater of Diamonds State Park.

was preferentially eroded down before cooling and settling back again. So the mystery remains, and Ozark hikers can ponder how it happened that our little piece of the Appalachian world became isolated from so much of that mountain system. This isolation is reflected, for example, in the fact that our region does not have some of the most common Appalachian tree species, such as the tulip poplar, and that our own species of chinquapin exhibits a DNA signature indicating that it is genetically distinct from its more well-known eastern relative. Whatever happened, it was not any minor effect, since more than a mile of subsidence would have been needed to create the required passage that separates our region from its eastern continuation.

With that background, let's look at some of the geological features that influence the things we see on a typical Ozark outing. One of the most characteristic features of our geology is the presence of chert (Figure 2.3), a glass-like mineral generated by silica deposited in the form of shells of diatoms, microscopic creatures that once lived in the upper ocean. The silica content of ocean water today is low because living organisms consume silica almost as fast as it arrives, such that oceans are often starved of this vital nutrient. In locations where water is fertilized with silica, abundant diatom shells settle into bottom sediments where they are buried, dissolved, and recongealed in clumps or layers as sediments solidify. The silica-bearing rocks that would make the Ozarks were deposited in a basin as the continent sagged under the weight of sediments eroding from the Ouachita Mountains, which were formerly much higher, in a manner analogous to that of the Appalachian-derived deposits in the Allegheny Plateau (Figure 2.1). Geologists have found evidence that the rate of sediment accumulation accelerated as the ancestral Ouachita range rose. During the intense mountain building of the Paleozoic era, active volcanos would have fertilized the basin to the north with a steady input of silica-rich volcanic ash, accounting for the great abundance of chert in some Ozark strata. Geologists argue today about whether that chert is derived primarily from diatoms or from volcanic deposition. This is largely an academic exercise, given that ocean water is starved for silica except where volcanic activity provided it. In other words, it's just a matter of whether that volcano-derived silica ever spent time as a diatom skeleton. But the chert issue is relevant to modern-day hikers,

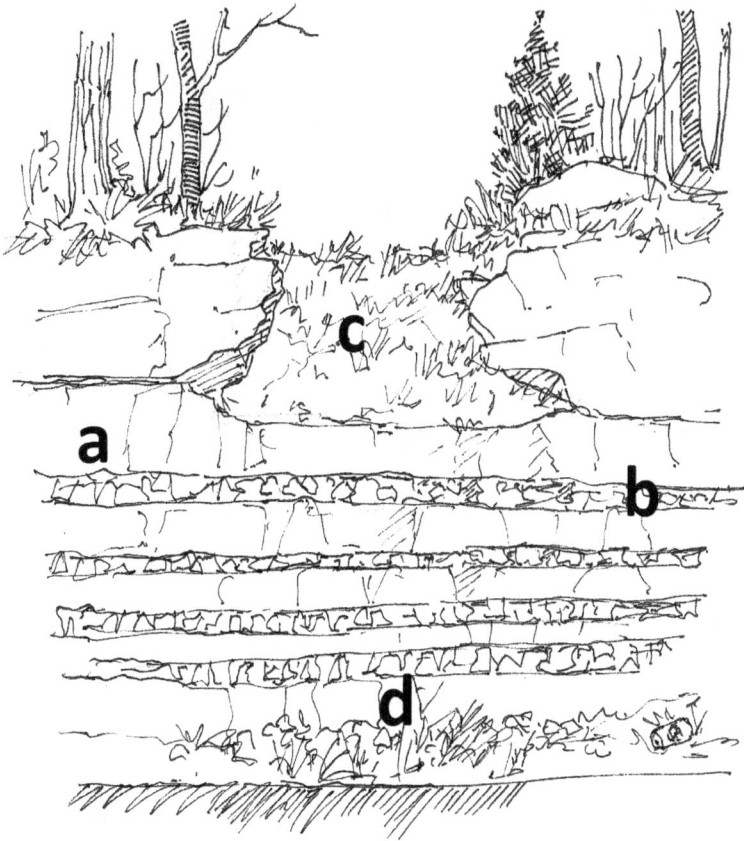

FIGURE 2.3. A road-cut in the Mississippian limestone that forms the surface of the Springfield Plateau, where thick gray limestone beds (**a**) are solid where cut into and contain tan chert beds (**b**). The limestone has developed cavities where groundwater has dissolved minerals in the shallow subsurface (**c**), while chunks of the brittle, easily fragmented chert become deposited at the bottom of the exposed rock face (**d**).

who often complain about those angular chert nodules, buried in fall leaf litter, that can twist ankles and trip the unsuspecting. Hikers will also notice sections of trail that seem to have been paved with white gravel, which actually represents the chert chunks that were left behind as the water-soluble limestone in which they were once embedded dissolved away (Figure 2.4). Now you know the unique process that has generated these familiar but sometimes puzzling features.

FIGURE 2.4. Limestone weathering on the surface of the Springfield Plateau leaves behind a residual gravel-like deposit that creates an acidic soil, often giving trails constructed in the Ozarks the appearance of having been paved with freshly crushed gravel.

Another characteristic of Ozark geology is the presence of ancient erosion surfaces separating many sedimentary layers, along with major lateral changes within individual units. All of this contributes to the confusion generated by consulting standard geological-sequence listings. For example, one often peers out from a mountain lookout to see exposed ledges running along the opposite side of the valley. Instead of forming one massive and continuous unit that you could identify on the chart, multiple bluffs or ledges vary in height and often seem to fade away altogether. Sometimes the bluffs can terminate on a fault where an adjacent block of rock has dropped down or been uplifted to break the continuity of the sediments. More often, we are seeing the effects of changing sedimentary environments along some kind of laterally varying environment (Figure 2.5). Think of a modern coastline and the many different environments you might see there, ranging from sandy beaches to muddy deltas and saline evaporating lagoons. Then think how these deposits might move across the landscape as sea level rises and falls over time. Figure 2.5 shows the

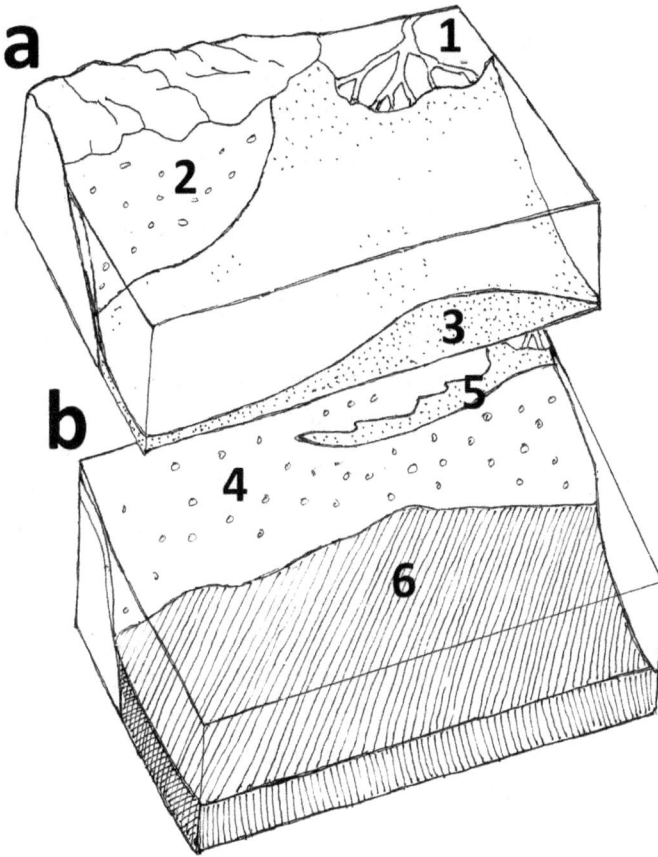

FIGURE 2.5. Schematic illustration of the Paleozoic coastal depositional environment as sea levels rose and receded over time. (a) Low-sea-level stand, where abundant sands from river deltas are fed directly into deep water (1), there is relatively
little shallow-water environment for carbonate deposition (2), and the thickness of
sand deposits depends on the distance to the river outlet source (3). (b) High-sea-
level stand, where there is extensive shallow water environment leading to widespread carbonate deposits (4), sand influx from river mouths is widely redistributed
by long-shore currents (5), and organic-rich shale is deposited in poorly oxygenated
deep water (6).

sequence of sediments and their variation along an ancient Ozark
coast during one of many cycles of sea-level change. You can see that
no single description is going to define this array of sediments, and the
difficulty is further compounded by periods in which this sediment
package is uplifted, warped, and eroded to various depths before the

next cycle of inundation. And this figure represents just a small part of a coastal environment with multiple deltas and embayments. The most rational approach to comprehending Ozark geology is to forget the nice textbook series of named formations and think instead of eons of oceans coming and going, and many subsequent ages of permanent exposure during which the slow passage of time has etched a rock-ribbed landscape, with the hardest layers projecting outward in a stair-step series of slopes.

The naturally layered structure of Ozark geology described above imposes a characteristic style of erosion. The hardest sandstone and limestone strata resist erosion, creating bluffs and cliffs. Other sediments, such as shale and siltstone, are more easily eroded to form the sloping benches between prominent ledges. Erosion of prominent sandstone cliffs proceeds by undermining blocks that slowly separate away from the main outcrop. Crevice caves are formed when blocks begin to separate from their cliff and slip to pile up on one another (Figure 2.6). The hardest rocks on hillsides also happen to be porous enough to allow downward seepage, either through pore spaces between sand grains or within crevices that inevitably develop in hard, brittle rocks. The softer shale and siltstone, however, prevent the passage of water downward while weathering to clay-rich sediments that can effectively lubricate the base of rock debris deposits. During times of high soil saturation, this lubrication process can lead to movement of large sections of slope, sometimes in slow motion—creating stilt-legged trees (Figure 2.7) and resulting in major landslides when roots alone cannot hold the land in place (Figure 2.8). These land-slippage events regularly affect Ozark hikers by requiring trail reroutes or, occasionally, even preventing the hiker from driving to the trailhead. The most spectacular erosion features are rock shelters, where openings develop as softer, lower layers of rock are removed from the base of a cliff line by the usual slope-movement process, landslide. Eventually, the sheer weight of the overhanging rock will bring a massive block of sandstone or limestone down to form another crevice cave, starting the process all over again.

One of the most interesting aspects of sandstone and limestone cliffs is the presence of a variety of intriguing features such as mineral textures and fossils, and these have always caught the attention of

FIGURE 2.6. Typical example of a crevice cave on an Ozark sandstone outcrop in the Boston Mountains where a large block of stone has begun to slip downhill, creating a cavity in the cliff face.

hikers in the Ozarks (Figures 2.9 and 2.10). Some texture is imposed on sandstones at the time of deposition. This is the pattern known as "cross-bedding," which appears as faint scallop-like lines within the stone. These result from bedforms (essentially underwater sand dunes) propagating down the sandy channel beds of a river delta. Sand grains are pushed up over the top of the bedform and cascade down the concave downstream side. You can often see the direction of flow in that ancient river from the shape of the cross-beds, as shown in Figure 2.9, a. Sandstone beds may also have undulating surfaces that

FIGURE 2.7. The exposed stilt roots of an elm show where rocks and soil on a shale bench have been periodically slipping or "creeping" downhill.

represent ripples on the surface of a sand body before it was buried under other sediments during deposition (Figure 2.9, c). Most other sandstone features are formed by the chemical processes that occur during consolidation as mineral-laden water flows through the consolidating sands under increasing temperature and pressure. One common result is the formation of concretions where crystalline minerals begin to coalesce around a seed point, creating a hard, ovoid or spherical, inclusion in otherwise uniform sandstone (Figure 2.9, d). These "seeds" can be small bits of organic matter that make a local change in the chemical environment. In more extreme examples, the mineralization can occur on an orthogonal network of seams or joints that form as the hardening sandstone contracts (Figure 2.9, b).

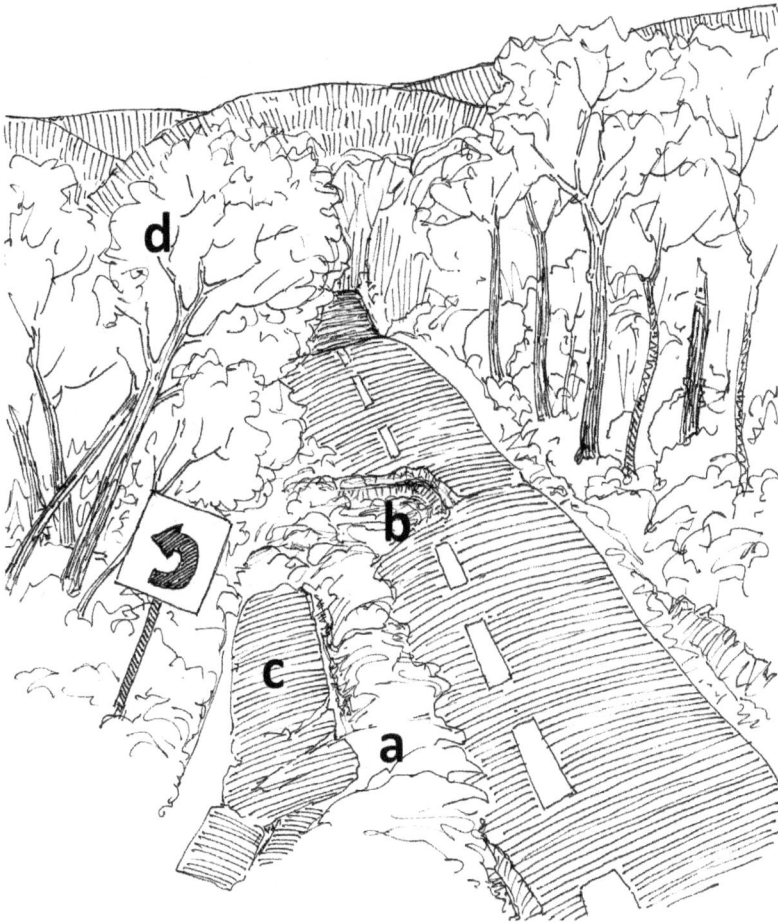

FIGURE 2.8. Disruption of the ground surface and damage to a road where an active landslide has recently occurred after a period of unusually heavy rain. The fluidized sediments have erupted to produce mounds of soil on the land surface (**a**), while the heavy pavement floating on this muddy fluid bed has fragmented under its own weight (**b**), with sections dropping down below the former roadbed (**c**). Rafts of soil made firm by tree roots also float on the fluidized mud, tilting at crazy angles (**d**) when the pressure subsides as fluid erupts onto the surface.

When such rocks are exposed, they develop an interesting latticework appearance because the mineralized seams have become more erosion resistant than the sandstone matrix in which they are embedded.

Most fossils are found in marine limestones in the Ozarks and represent creatures such as brachiopods (Figure 2.10, a), crinoids (Figure

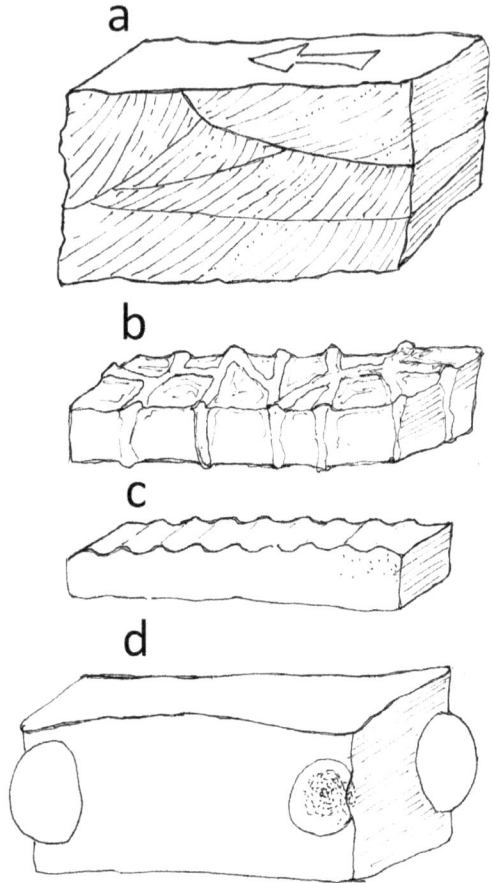

FIGURE 2.9. Typical sandstone rock textures seen in the Ozarks: (**a**) cross-bedding where the scalloped pattern in the rock is caused by sand grains deposited on the downstream side of dune-like bedforms in rivers that show the direction of river flow; (**b**) lattice-like patterns created when erosion-resistant quartz is deposited in the joints within the sandstone long after deposition; (**c**) ripples created by gentle flow over a deformable sand bed that is then buried by additional sand; and (**d**) cannonball-like concretions embedded in massive sandstone or the round sockets where such concretions have fallen out of the rock face.

2.10, b), and corals living in the shallow seas that sometimes inundated the area. Plant fossils are much less common in Ozark sandstones and usually represent some of the most common fossil plants, such as the giant lycopods and seed ferns of the vast coal swamps that were located farther inland (Figure 2.10, c). Many of these fossils represent driftwood brought downstream into the largely unforested, sandy deltaic deposits that would become part of the Ozark uplift dome.

Meanwhile, we normally think of coal as being a product of regions east of the Ozarks. However, there are some relatively thin layers of coal in the Boston Mountains and other Ozark locations (Figure 2.11). The Ozarks served as a shoal area in the shallow seas to the west of the region where great deposits of sediment, including

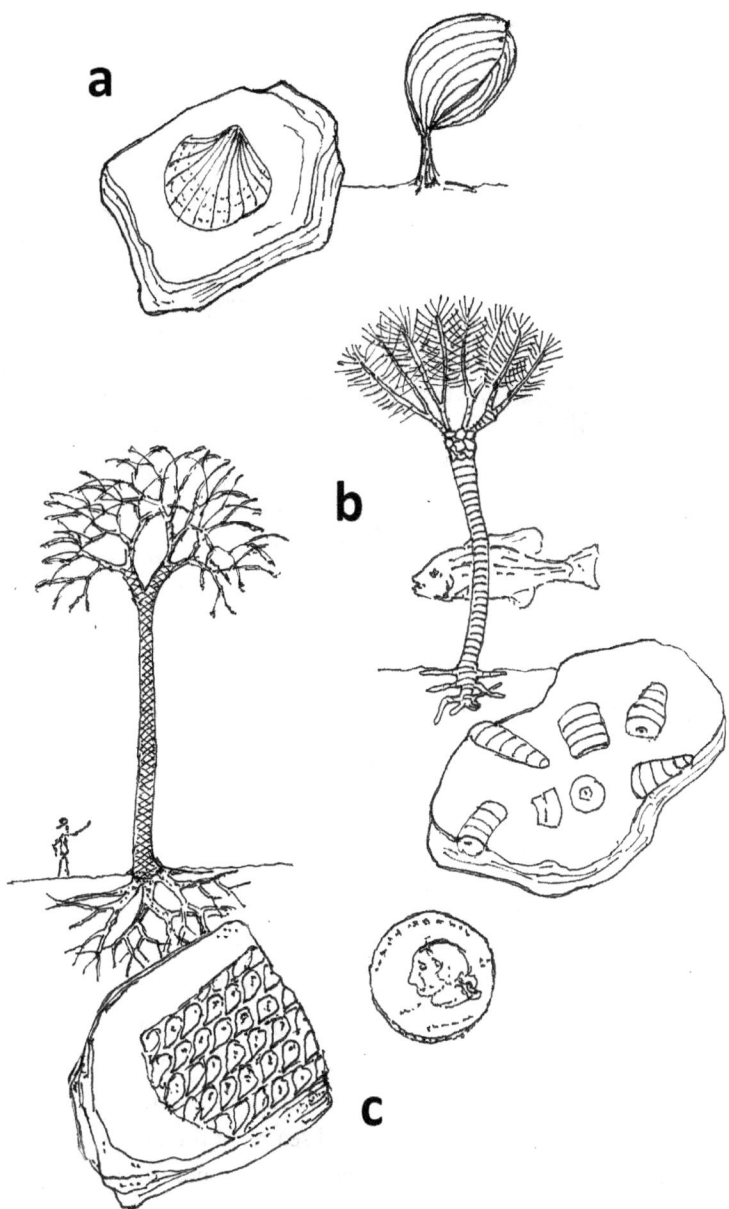

FIGURE 2.10. Three of the most common fossils found in Ozark rocks, shown beside the life forms that created them: (**a**) brachiopod fossil shell in limestone; (**b**) crinoid (or "sea lily") stem fragments in limestone; and (**c**) imprint of the scaly bark of a lycopod (*Lepidodendron*) in sandstone. A quarter provides the scale for all three fossils, while a small fish and a human figure give the scale for the intact crinoid and lycopod, respectively.

FIGURE 2.11. Thin coal seam over thin shale, typical of Ozark outcrops; the original thickness of the peat deposit that was compressed to form the coal is estimated to be about ten times the final coal-bed thickness, so these thin deposits were originally five or ten feet thick (analogous to the peat layers now found in the Florida Everglades).

coal, were being emplaced in subsiding basins at the foot of the high Appalachians. Some of the rivers carrying those sediments then traversed the Ozark region, generating some of the sandstone layers we see today. This is not a conjecture; geochemistry and radioisotope dating of minerals (erosion-resistant zircons) in the sandstone confirm this origin. Geologists figure that there is a ten-to-one compaction of coal as it matures, so that the thin coal seams of the Ozark region that are a fraction of a foot thick today were once similar to the eight-foot thickness of peat in the center of the Everglades slough (a modern analogue to our ancient landscape). And the story of coal provides a fascinating insight into the past of Ozark forests. As discussed in a subsequent chapter, the decay of forest litter and recycling of nutrients is closely coupled with the activity of fungi living within the soil. Scientists believe that the huge deposits of coal that date to the so-called "Coal Age" resulted from the inability of the primitive fungi that existed at the time to break down the tough lignin fibers in the woody tissue of plants. But that's not all. The deposition of so much carbon in sediments caused the oxygen left behind to accumulate in the atmosphere. Today, we have about 20 percent oxygen in the air we breathe. Back then, oxygen may have been as high as 35 percent. Imagine the raging forest fires that must have existed at that time to consume every stick of wood on uplands and burn their way deeply into even the wettest swamp deposits. You might even think of the lowly mushroom you see on your walk in the woods as nature's way of fire-proofing the forest.

The presence of several major water-soluble limestone units in the geology of the Ozarks results in a special kind of terrain known as "karst." Solution enlarges cracks in limestone layers, allowing the development of an underground drainage system (Figures 2.12 and 2.13). This underground plumbing is seen in bedrock exposures as solution-enlarged vertical openings in rock walls created along roads (Figure 2.12A) or in the form of sinkholes where overlying sediments have collapsed into these openings from above (Figure 2.12B). Even when there are no vertical joints along which subsurface solution activity can occur, the seepage of groundwater laterally along the buried rock surface generates a complex network of channels known to geologists as "epikarst" (Figure 2.13). In many other regions, streamflow slowly declines with time after a rainfall event as water that has seeped into the porous soil and gravel in the streambed slowly flows back into the stream channel, a process hydrologists call "baseflow." Many karst-region streams do not exhibit such baseflow because only the excess that cannot be conducted by the underground flow system will flow in the aboveground channel, and only during extreme rainfall events. The subsurface flow returns to the surface in springs (see Chapter 8). In many cases, the underground drainage area for the karst system is completely unrelated to the surface drainage system that lies above, and that notoriously affects considerations related to identifying possible sources of contamination, such as campsites and outhouses, that park managers need to consider.

Although most of Ozark geology involves sedimentary rocks,

FIGURE 2.12A. Example of karst weathering where solution openings created by groundwater flowing downward along vertical joints in the rock are exposed by a recent road-cut in Mississippian carbonate sediments at a site on the Springfield Plateau in northwest Arkansas.

there is an exposure of the underlying "basement" rocks in the center of the dome in southeastern Missouri (Figure 2.14). These are much more ancient crystalline metamorphic rocks that impose a different texture to the landscape that is somewhat analogous to the Blue Ridge Mountains, which are underlain by the same kind of old metamorphic rocks. A good place to experience this landscape is at Elephant Rocks State Park in southeastern Missouri. Most visitors might wonder where all these "elephants" came from. They appear as rounded boulders almost randomly dropped down on exposed pavements of solid rock. We think of elephants as living in the jungle, and that is probably where these pachyderms originated. The landscape in Figure 2.14 looks so different from what we see elsewhere in the region because the process that produced them was far different from those active today. These kinds of boulders originate by chemical weathering beneath deep tropical soils under the action of acidic

FIGURE 2.12B. Typical Ozark mountain sinkhole where overlying sediments have collapsed into a solution-enlarged vertical joint in a limestone formation situated directly beneath this location.

water infiltrating from above (Figure 2.15). The acidity comes from infiltration through organic litter in a heavily forested landscape. As solid bedrock is uplifted, the rocks develop a rectilinear pattern of joints from stress relief. Water penetrates this network and dissolves the rock, working inward. The result is a set of rounded boulders suspended in weathered clay-rich residue. If further uplift occurs and that residue is washed away, the elephant rocks are deposited on their perches. In *National Geographic* photos, you often see lions sitting on similar piles of exposed, tropically weathered boulders, known as "kopjes" in South Africa. Try to imagine the proud lions that once inhabited our continent standing on our elephant rocks, when those

FIGURE 2.13. The formerly buried surface of limestone, exposed by diversion of a stream over a reservoir spillway in northwest Arkansas, shows the irregular surface created through dissolution by groundwater seepage that makes it look as if the bedrock has been literally melting to form what geologists call "epikarst."

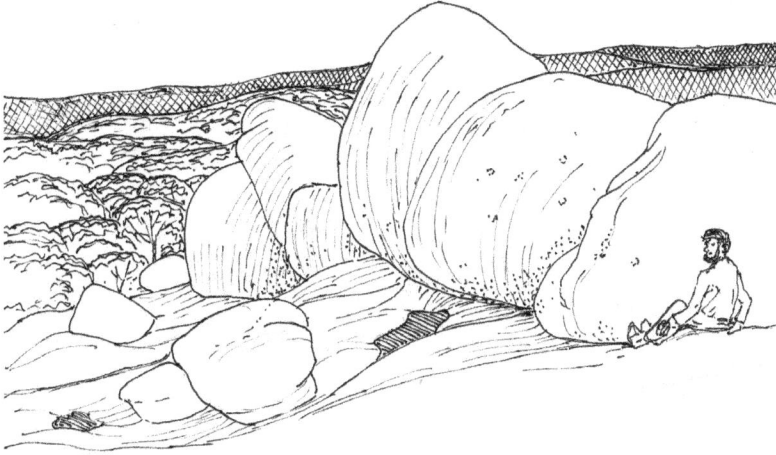

FIGURE 2.14. Boulders developed on a bedrock exposure at Elephant Rocks State Park in Missouri, where granitic basement is exposed at the surface along the top of the Ozark dome.

regal cats still patrolled the Ozarks. On a geologic time scale, this was not long ago at all.

One additional consideration related to Ozark geology is the effect on the landscape of change imposed by climate and land use. Although our short life span does not allow us to easily recognize the fact, conditions on Earth are rapidly (on the geologic time scale) changing over time, and the plant and animal inhabitants of the Ozarks have had to cope with those changes during their evolution. Extremely slow changes are associated with the way that positions of continents affect the global energy balance—changes that occur on the scale of one hundred million years at a time. Much more relevant are shorter-term oscillations in the global energy budget caused by what would seem insignificant variations in the way that different regions receive sunlight. When Serbian mathematician Milutin Milanković (also spelled "Milankovitch") originally proposed this effect as the driver behind the climate fluctuations of the Ice Age, his theory was dismissed out of hand. The details of how this mechanism was confirmed as the Earth's climate driver are beyond the scope of this book. The relevant

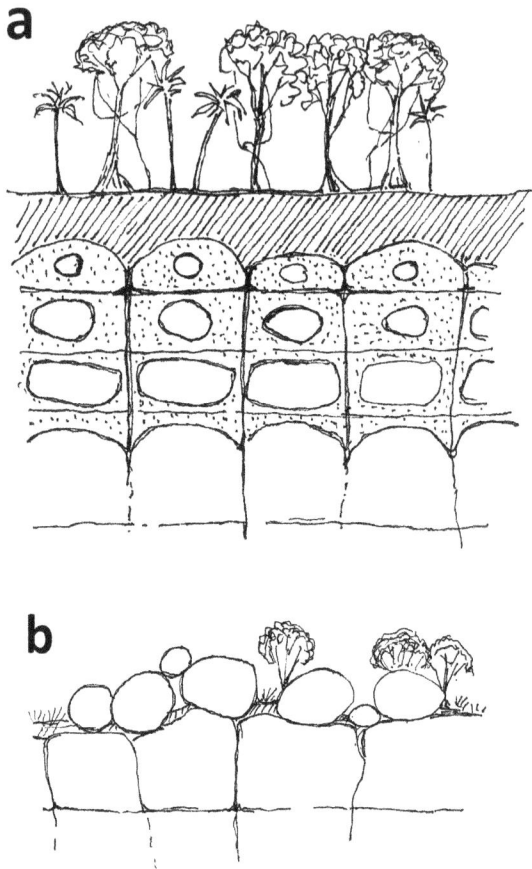

FIGURE 2.15. Schematic illustration of the formation of Elephant Rocks. (a) Stress relief from uplift forms rectangular joints in deeply buried bedrock, such that acidic groundwater seeping down from a tropical rainforest can chemically soften the outer surface of the blocks of rock. (b) A change in climate leads to mechanical erosion and exposure of large, rounded blocks of rock that once formed the cores of chemically weathered bedrock blocks.

fact is that the Earth has an elliptical orbit such that the time when our Northern Hemisphere has its closest approach to the sun in the winter recurs on a twenty-thousand-year cycle. This produces times when summers are unusually cool and when snow in the far north doesn't melt completely. Snow builds up and reflects more sunlight, providing more cooling even when the summers have warmed up again halfway through the cycle. There are pulses of ice buildup at twenty-thousand-year intervals until there is finally a summer warm enough to make the ice sheets collapse. Then the whole process starts up again, in a series of endless cycles that extends back through all of the time represented by Ozark sediments.

The Ozarks are located beyond the reach of the most recent glacial advances, but the relentless pulse of the "Milankovitch cycles" extends around the world, the coldest glacial times near the poles corresponding with arid times over the rest of the globe. The ecological associations we see today in the Ozarks have been shaped by this essentially infinite past cycling of climate change through the eons of geologic history. The impact of these climate changes can be assessed using fossil pollen taken from the mud at the bottom of water bodies, providing a detailed inventory of the plants and trees adjacent to a lake or pond. The best source of information for the Ozarks is from Cupola Pond in southeast Missouri, which contains nearly twenty thousand years of sediments according to radiocarbon dating methods (Figure 2.16). The change in pollen percentages over time at Cupola Pond, combined with extrapolation from pollen and macrofossils recovered at more distant sites, suggests a series of changes over time as illustrated in Figure 2.17. During the coldest times prior to sixteen

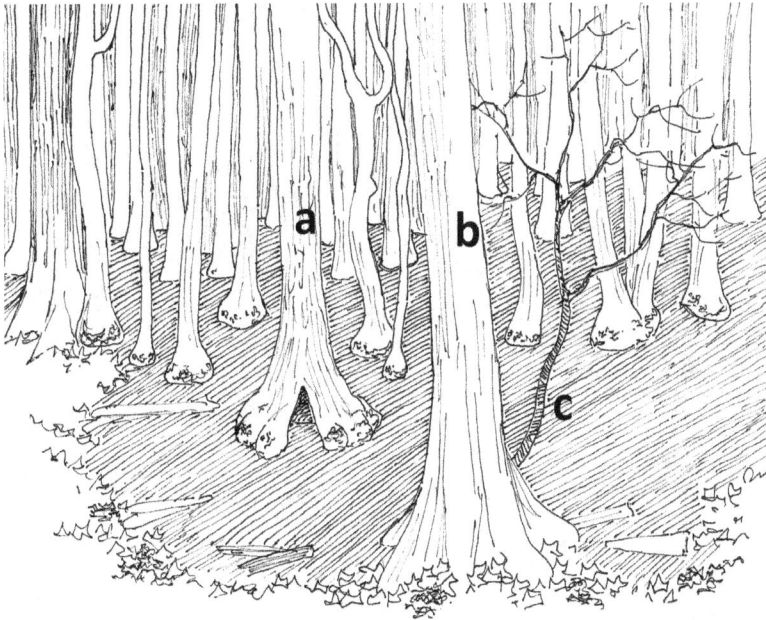

FIGURE 2.16. Cupola Pond in southeastern Missouri, as it appears today: a shallow, acre-sized depression filled with water gum trees (**a**) and large pin oaks (**b**) around the edges, with a few stunted red maples (**c**) straddling the roots of the gums and oaks.

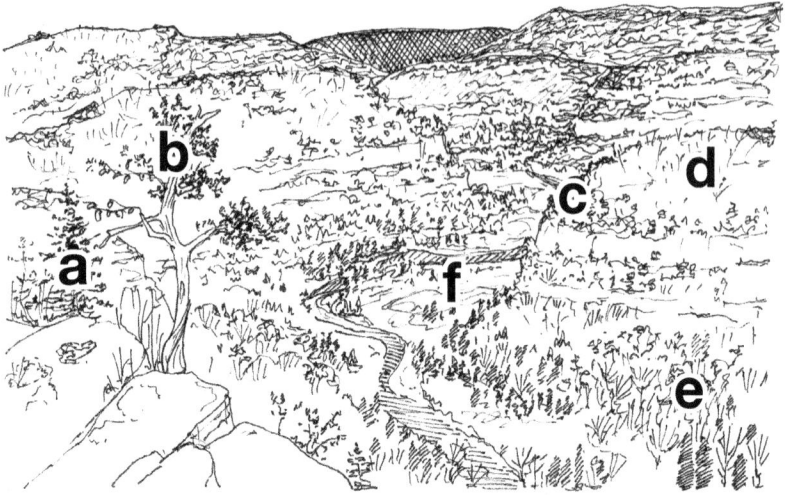

FIGURE 2.17A. An Ozark mountain scene of eighteen thousand years ago, during the height of the last glaciation. The dominant trees are spruce (**a**) and jack pine (**b**), with the uplands covered by a relatively open pine forest (**c**) subject to extensive fires that leave open, burned-over areas (**d**). Sheltered locations have a forest of mature spruce mixed with bur and northern red oaks (**e**). Small prairies exist along river bottoms (**f**).

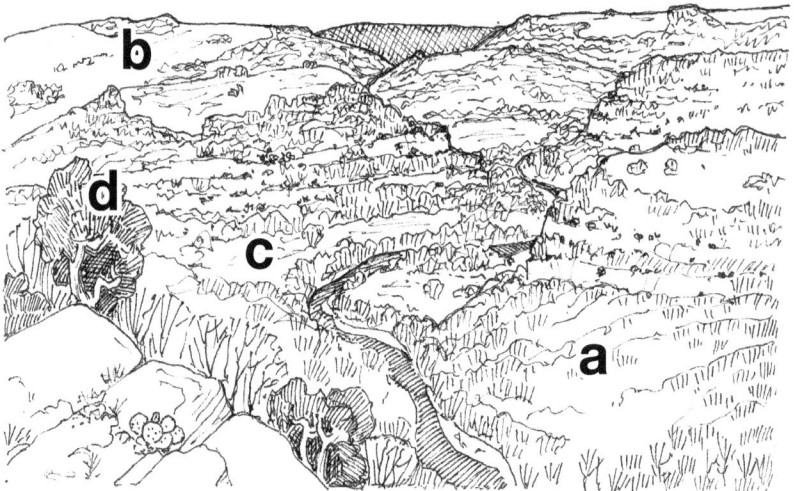

FIGURE 2.17B. An Ozark mountain scene of six thousand years ago. The landscape is covered by an oak-dominated forest (**a**) with extensive open areas on plateau uplands (**b**) and tallgrass prairies along river bottoms (**c**). There is no pine present, and the only conifers are limited numbers of red cedars restricted to rock ledges (**d**).

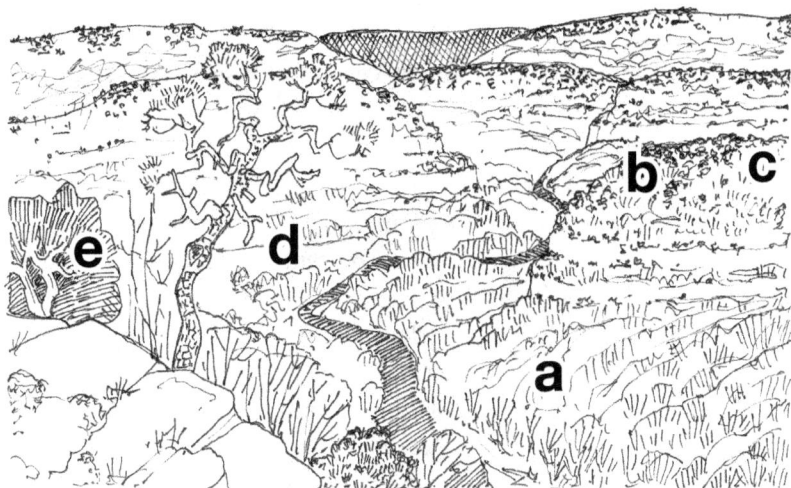

FIGURE 2.17C. An Ozark mountain scene in about 1500 A.D., before European settlement of the area. Oak–hickory deciduous forest covers slopes (**a**), with pine on the uplands (**b**), and a distinct line of demarcation separates the two kinds of forest (as described by early explorers Dunbar and Hunter). There are burned-over areas in the pine forest (**c**) and small tallgrass prairies along river bottoms (**d**). Red cedar is restricted to the immediate vicinity of rock ledges (**e**).

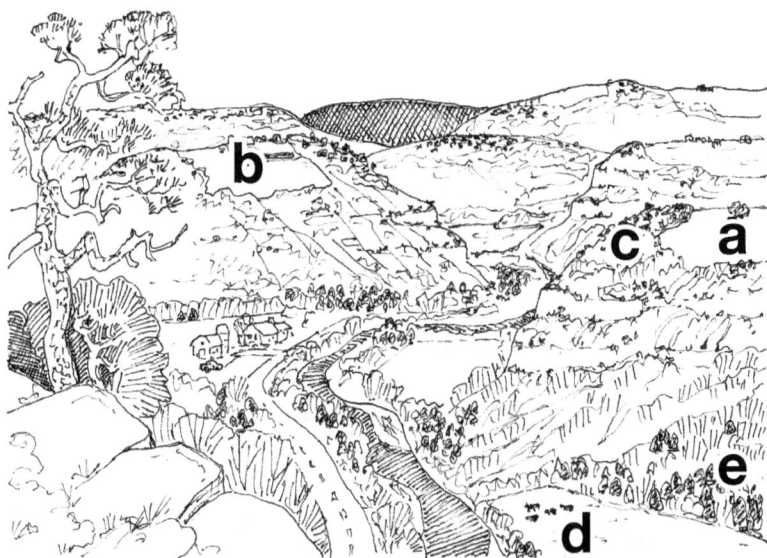

FIGURE 2.17D. An Ozark mountain scene today. Large areas on the plateau uplands are converted to pasture (**a**) with numerous livestock ponds (**b**). Pine is still a major component of the forest on uplands (**c**), and most of the bottomlands have been converted to cultivated fields or pasture (**d**). No longer limited to rock ledges, red cedars have created pure stands on abandoned fields and are now a major component of disturbed woodlands (**e**).

thousand years ago, the Ozarks had a relatively open landscape in which spruce and pine predominated, but some of the more cold- and drought-tolerant oaks were almost certainly present (Figure 2.17A). A good analogue would be southeastern Manitoba, where the pine species is jack pine (Figure 2.18) and where bur oak grows today. In fact, fossil jack pine needles from another site in eastern Kansas confirm the pine identification for our region. Ironwood and perhaps northern red oak were also probably present. As glaciers to the north faded away and the Ozarks warmed, the climate became somewhat warmer and dryer than it is today (Figure 2.17B), during a period often referred to as the Altithermal (spanning nine thousand to six thousand years before present). The Ozark landscape would have been similar to that of today except for larger prairie and grassland openings on well-drained sites, significantly more open forests, and the complete absence of pine. Just before the earliest records from European visitors to the Ozark area, humans would have already been present for some time. Their preagricultural impact would have been negligible, except for a probable role in the extinction of megafauna such as the mammoth and horse. Agriculture was in use on a limited scale in late prehistoric times (ca. 1500 A.D.; Figure 2.17C), and surrounding forests would have been somewhat more open than today's woodland. Finally, we see the settled and developed Ozark landscape of today (Figure 2.17D), with pastures, cultivated fields, and a somewhat more closed woodland in adjacent forests as a legacy of more than a century of fire exclusion (see Chapter 5).

This brief survey of the geological history of the Ozarks is intended to instill an appreciation for the many factors that created the landscape we enjoy today. Simplified layer-cake descriptions of sedimentary columns can hardly do justice to the amazing series of events that have given us our scenic mountains and rivers, regionally isolated from their counterparts in the greater Appalachian region. Outdoor enthusiasts should take solace in the fact that the landscape we have in the Ozarks is a whole lot more interesting than the simple set of layers portrayed in the diagrams of standard geology textbooks.

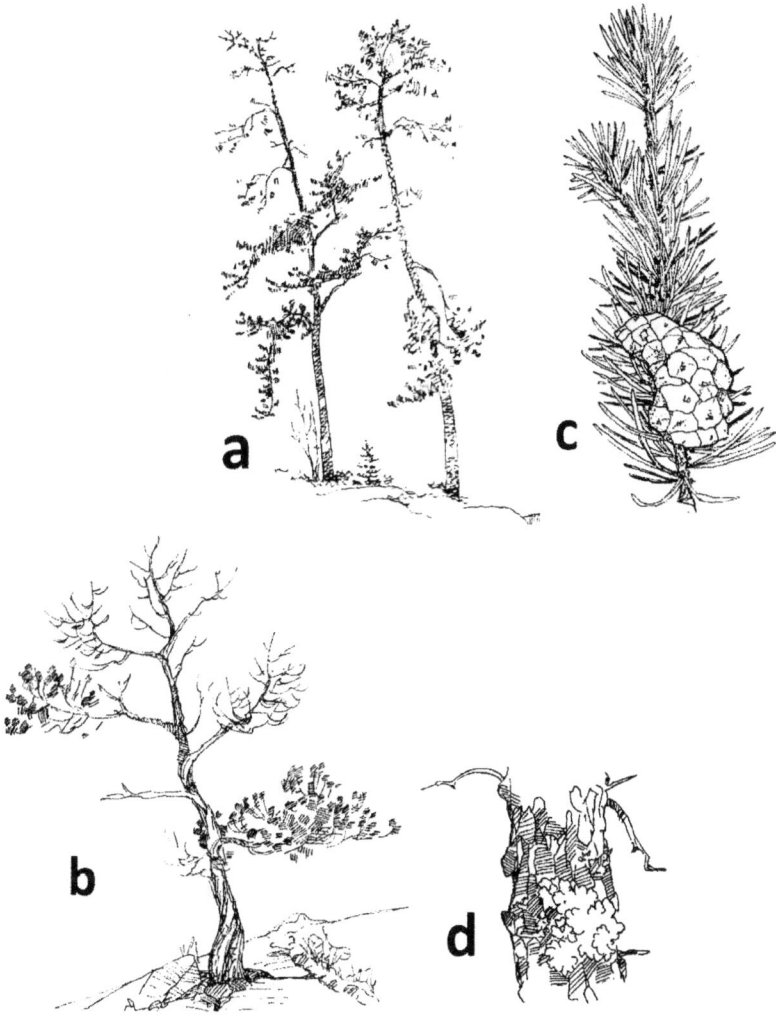

FIGURE 2.18. Jack pines growing on and around bedrock outcrops at the prairie–forest border in southeastern Manitoba today: (**a**) typical woodland specimens (note stature and shape), (**b**) a stunted individual on a rock ledge, (**c**) detail of needles and cone, and (**d**) lichen-encrusted bark. Jack pines were probably growing in this form and under a similar climatic regime on rock outcrops in the Ozarks eighteen thousand years ago.

CHAPTER 3

The Principal Trees in an Ozark Forest

A forest is defined on the basis of the major types of trees that make it up, as delineated in a host of U.S. Forest Service technical publications. For example, the forests of the Ozarks are generally designated as "oak–hickory forest," since various species of oak and hickory are the most consistently widespread and abundant trees present—though forests containing quite different assemblages of species occur in some situations, especially along streams. Oak–hickory forests are not limited to Arkansas; they occur throughout a broad region that extends from southern New England to northern Georgia and as far west as northeastern Texas and southern North Dakota. The deciduous forests of southeastern North America are known for their diversity of tree species, and the many varieties of oak and hickory that range into the Ozarks can be confusing for all but the most knowledgeable specialists. The nine species of oak and five species of hickory that are most common and widespread in our area are illustrated in Figures 3.1 and 3.2 (and the characteristics of these oak species are summarized in Table 3.1). A few other hickory species occur along the edges of the Ozark region, and a number of other oak species can also be found, including the scarlet oak, whose range includes the northeastern part of the Ozarks, and the bur oak, which occurs locally in small populations. But a hiker who is familiar with the fourteen trees illustrated in Figures 3.1 and 3.2, along with a few other common species (Figure 3.3), will recognize most of the principal actors in the forests that he or she encounters. Famed nuclear physicist Edward Teller once denigrated biologists, calling their classification of species about as technically difficult as stamp collecting. But there is a whole lot more to the process of tree identification than just species identification, and

TABLE 3.1

Practical identification keys for the most common species of oaks in the Ozarks and the relation of each species to specific mountain habitats.

SPECIES	LEAF	BARK	PREFERRED HABITAT
Northern red	Broad, with moderate depth to lobes and symmetric shape	Dark gray with long linear ridges that often appear silvery	Relatively rich and moist sites such as small streams and bases of ledges
Black	Variable depth to lobes and non-symmetric shape; leaf shape varies greatly	Dark gray ridges usually broken into rough rectangular plates; inner bark has a distinctive orange color	Upper slopes and dry ridgetops
Southern red	Often three main lobes and "turkey track" shape	Similar to black oak except the plates are smaller, more square than rectangular	Well-drained south-facing slopes and some stream bottoms
Shumard	Broad, symmetrical; three lobes of nearly the same width	Almost identical to northern red oak, with vertical bark ridges	Alluvial soils of larger streams
Pin	Deeply lobed and symmetrical leaves with sharp "pin" points	Moderately furrowed with medium gray color	Poorly drained depressions and flood plains
Blackjack	Triangular shape with shallow lobes and very short leaf stem	Dark gray, nearly black, with rough pebbly texture	Driest slopes with poor soil, especially shale
White	Deeply lobed leaves with rounded ends to lobes	Light gray with rough rectangular bark plates where some seem to be flaking off	Deep soil on middle and upper slopes

TABLE 3.1 (CONTINUED)

Post	Three main rounded lobes give the leaf a cruciform shape	Light gray bark broken into small rectangular plates arranged in spiral pattern on mature trees	Dry upland sites or edges of prairie openings
Chinquapin	Elongated oval shape often broader toward tip and sawtooth edges	Light gray flakes that are more rounded than rectangular	Variety of sites with high-calcium soils

we will show how much can be learned about the landscape around us, including its history, just by observing which trees are present and in what proportions.

Leaves are a good place to start in considering our forest trees. These energy-collecting solar cells are a critical part of any tree's structure. Indeed, a basic function of the tree is to position its leaves to effectively intercept sunlight. Despite the apparent simplicity of that function, we find that the leaves of trees vary considerably in size, shape, and structure. They also differ in leaf arrangement—how they are attached to the twigs from which they arise. In the majority of trees found in Ozark forests, the leaf arrangement is "alternate." This means that only a single leaf arises at a particular level on the twig, with the next (or previous) leaf being attached at a higher or lower level and on the other side of the twig. This is the leaf arrangement in, for example, oaks and other nut-bearing species (Figure 3.4, a). But there are other tree species on which the leaves arise in pairs, with one leaf in each pair attached directly opposite the other. This "opposite" leaf arrangement is characteristic, for example, of maples (Figure 3.4, b).

Most commonly, a leaf consists of a single, flat, expanded portion ("blade") that arises from a stem-like portion ("petiole") that is attached to the twig. This type of leaf is referred to as "simple." However,

| Northern red oak | Black oak | Southern red oak |

| Shumard oak | Pin oak | Blackjack oak |

  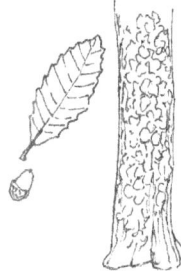

| White oak | Post oak | Chinquapin oak |

FIGURE 3.1. The nine most common and widespread species of oak in the Ozark region, showing the relative size and shape of leaves and acorns and typical bark textures.

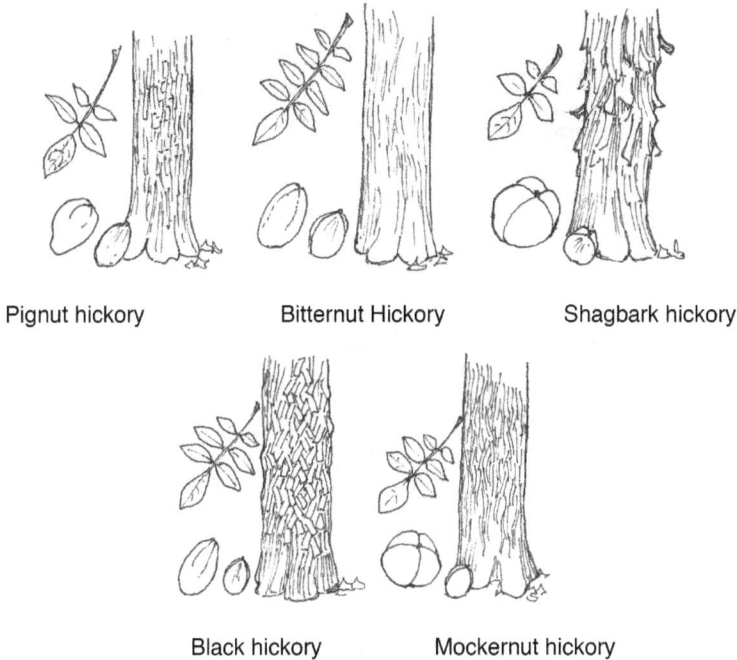

Pignut hickory Bitternut Hickory Shagbark hickory

Black hickory Mockernut hickory

FIGURE 3.2. Leaves, nut husks, and nuts of the five most common and widespread species of hickories in the Ozark region, along with the bark pattern and texture for each species; leaflet numbers can vary within each species (the most common leaflet number for each is shown). Note the variation in the degree of angularity on nuts, ranging from perfectly round in cross section (pignut) to sharply ribbed (shagbark).

in some trees, the blade is divided into a number of smaller, leaf-like portions ("leaflets") that can be confused with true leaves unless one examines them closely or already knows that the tree in question has what are referred to as "compound" leaves (Figure 3.4, c). Hickories (Figure 3.2) are among the more common examples of trees characterized by compound leaves, but this group also includes walnuts, white ash, and black locust.

The blade of a leaf or leaflet may have a smooth margin (referred to as "entire") or it may have lobes or teeth. In a leaf with lobes, the margin of the blade is divided into a number of rounded or angular projections that are usually rather conspicuous. A leaf is said to have "teeth" if there is a series of small notches (often like the edge of a saw)

Black gum

Black walnut

White ash

Winged elm

Black cherry

Sugar maple

Red maple

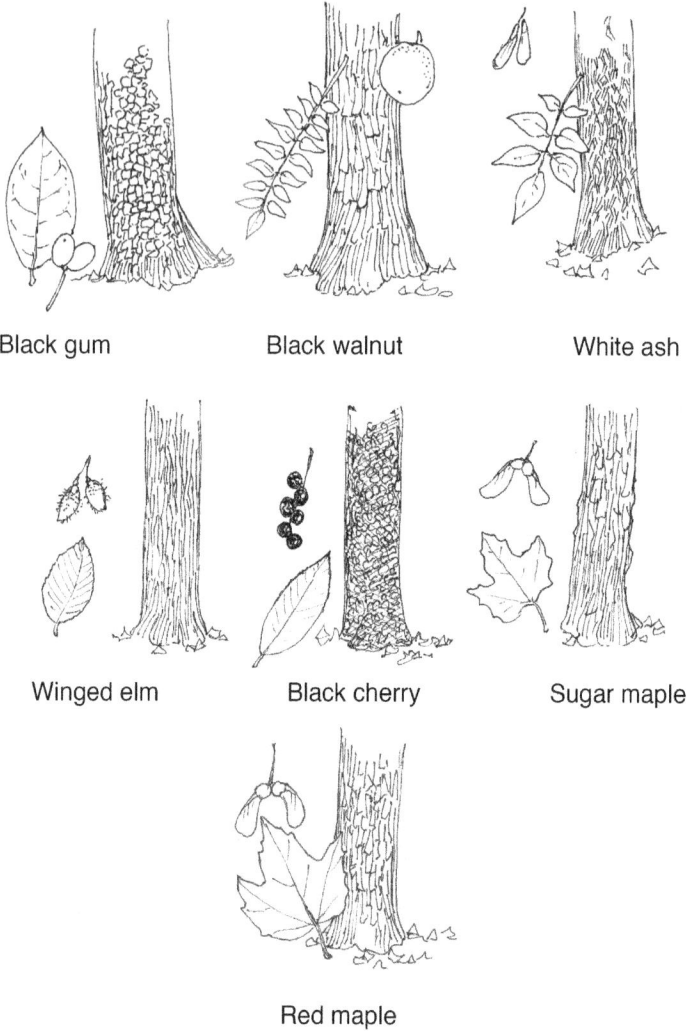

FIGURE 3.3. Leaves, fruits, and bark textures of other common Ozark forest trees.

along its margin. Teeth may occur only at the base of the blade, be concentrated toward the tip, or (most commonly) occur along the whole margin (Figure 3.4, a). They can vary considerably in size and number, and in some instances it's necessary to examine the leaf closely in order to detect their presence (Figure 3.4, c).

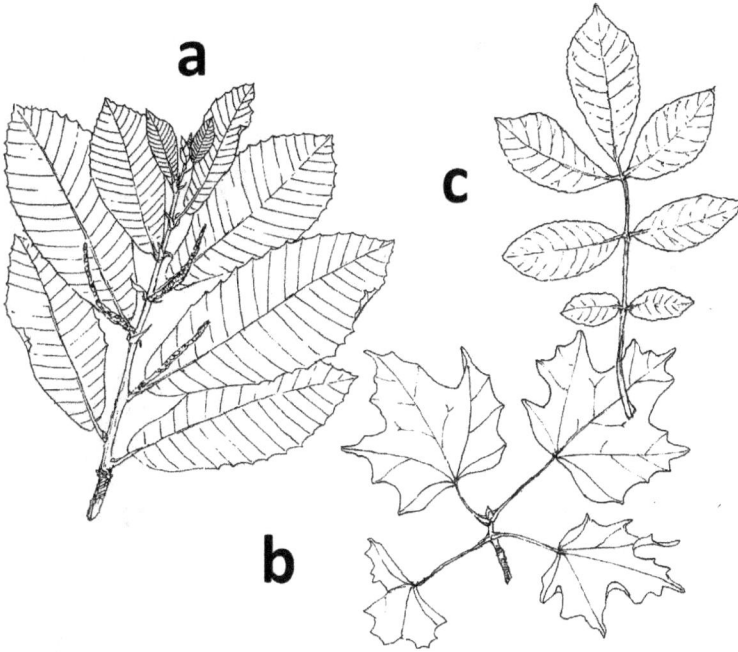

FIGURE 3.4. Leaf-arrangement geometry in deciduous trees: (**a**) alternate leaf arrangement on a twig of Ozark chinquapin; (**b**) opposite leaf arrangement on sugar maple twig; and (**c**) multiple leaflets making up the compound leaf of mockernut hickory.

Although we usually think of identifying trees by one characteristic leaf shape, that is often not the case. Most trees can modify the structure of their leaves for the specific environment in which they are located. Thus, a tree will typically have "shade leaves," which are broader, thinner, and more efficient in low light; and "sun leaves," which are narrower, thicker, and have a tougher outer cell layer so that they can withstand exposure to full sun and wind (Figure 3.5). This occurs because leaves have to balance the net gain of energy produced by photosynthesis and carbon dioxide uptake with the loss of water to the surrounding atmosphere. That balance will be different in the shaded environment below than in the exposed position at the top of the tree. Hikers therefore have a biased impression of the trees around them, seeing mostly understory shade leaves. If you are walking in the forest after a summer thunderstorm, take the time to look for

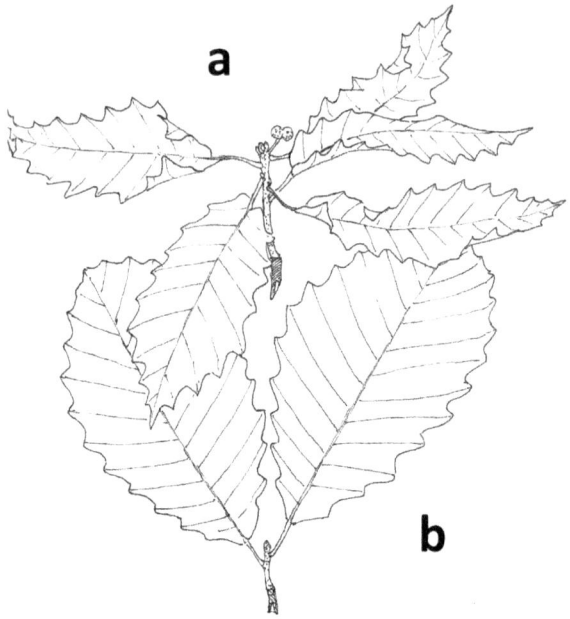

FIGURE 3.5.
Comparison of
(a) "sun leaves" and
(b) "shade leaves"
of chinquapin
oak. Sun leaves are
generally smaller
and thicker, with
curled-up edges,
while shade leaves
tend to be larger,
thinner, and almost
perfectly flat.

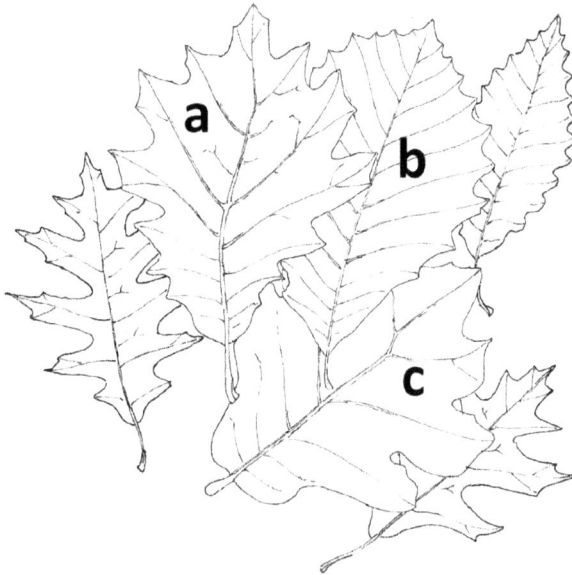

FIGURE 3.6. Oversized leaves on rapidly expanding sprouts from the bases of oak saplings injured by fire, or from the stumps left when trees are cut, are typically twice the size and very different in shape from those on the branches of mature trees. Compare the juvenile leaves of (a) northern red oak, (b) chinquapin oak, and (c) black oak with the mature leaves of these species shown underneath each of these examples.

small, fallen branches from high in a tree's crown to get a close look at sun leaves and compare them to the leaves of the same species in the understory. These differences complicate leaf recognition, which is even more difficult in the case of rapidly expanding shoots from stumps and small saplings, something the hiker is likely to see because these vigorous shoots are present alongside the trail. The difference between mature and juvenile leaves is especially great in some of the oaks and can cause serious confusion when you try to figure out what species you are looking at (Figure 3.6).

One other feature useful in identifying trees is leaf "venation," a term that refers to the way in which the main veins in the blade of the leaf (or the leaflets if the leaf is compound) are organized. In the "palmate" type of venation, the main veins (or leaflets) all arise from one point at the base of the blade and radiate outward like the fingers of a hand (Figure 3.4, b). In the "pinnate" type, the main veins arise at regular intervals along the midrib of the blade (Figure 3.4, a). The most commonly encountered trees that produce leaves with palmate venation are the maples, while such trees as oaks have pinnate venation.

Their function as solar-energy-collection surfaces influences the way leaves are arranged in the tree canopy. Moreover, trees are rated as "tolerant" or "intolerant," depending on the energy efficiency of their leaf surfaces. Intolerant trees generally have leaves that are most efficient in full sunlight and therefore do not "tolerate" shade very well. Tolerant trees have leaves that are optimized for lower light levels but are not as efficient at high light intensity as some other trees. As a general rule, intolerant and mid-tolerant trees such as oak and hickory have a branching pattern that is relatively open, allowing interception of light by leaves arrayed in multiple layers (Figure 3.7). Tolerant trees such as maple tend to have a single, dense layer of leaves arrayed along the top of the tree's crown. The lower branches of both classes of trees tend to die back (a process known as "self-pruning") because even the shade-tolerant leaves of maples lie beneath the dense leaf layer produced by the maple's upper branches.

Another informative aspect of the leaves of forest trees is the relationship between their size, shape, and degree of indentation. As a general rule, the broader the leaf and the shallower the lobes or indentations on the leaf, the more favorable the climate in which the tree grows. Paleontologists have created an elaborate rating system for

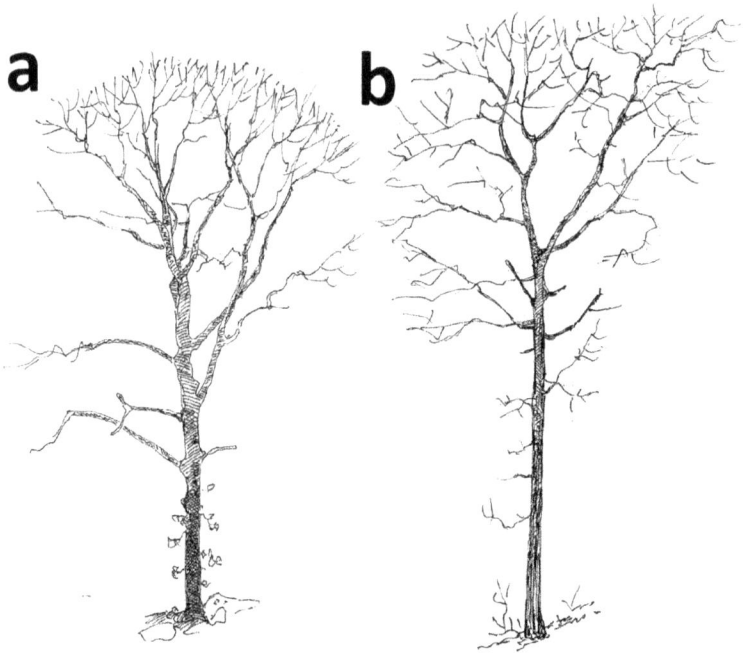

FIGURE 3.7. Comparison of crown structure: (**a**) tolerant trees such as this sugar maple develop a single, dense leaf layer across the top of the tree; whereas (**b**) intolerant trees such as this black oak develop a diffuse crown of more dispersed leaf layers.

the shapes of fossil leaves from trees that lived in ancient eras, which has been used to construct quantitative estimates of temperature and humidity in landscapes that vanished long ago. This is illustrated nicely by comparing the leaves of the three oak trees in the top row of Figure 3.1. Of all the red oaks in the Ozarks, the northern red oak inhabits the most mesic sites ("mesic" refers to the most fertile and moist habitats). That species has the largest and broadest leaves of all our red oaks. By contrast, the southern red oak has smaller, much more deeply indented leaves and inhabits relatively dry habitats on south-facing slopes. Black oak has intermediate leaves and inhabits a whole range of sites in between those most favorable for the two red oaks. Even so, we have regularly noted southern red oaks growing in the rich and moist soils of small stream terraces in the Ozarks. This may hint at a question of multiple identities in this species: taxonomic textbooks identify two separate races of southern red oak, one the familiar tree of dry and impoverished uplands, the other (known as "cherrybark oak") a massive tree of Mississippi River bottomlands. Perhaps our own southern red oak is expressing this known form of arboreal identity confusion.

Besides leaves, the other major identifying component of the forest with which the hiker is intimately familiar is the tree's trunk. As indicated in Figures 3.1, 3.2, and 3.3, the physical appearance of bark is one of the characteristics used to differentiate various species from each other, especially in winter when no leaves are available for reference. Much of the wood in the trunk of a tree simply serves as mechanical support for its branches and cells that once transmitted water and nutrients upward but are now filled with resin and other substances to form what we call "heartwood." The actual living tissue of wood and bark is contained within an outer cylinder of conductive cells, in which a double layer of living tissue provides the cellular transport tubes that conduct fluid upward on the inside ("xylem") and downward on the outside ("phloem") (Figure 3.8). Outside of these two layers is what we call "bark," and our interest is concentrated on this visible part of the tree trunk. As in the case of leaves, it is a wonder that a part of tree anatomy with such a simple and straightforward function should display such a wide variety of colors, thicknesses, and textures. The bark consists of dead cells that serve to protect and insulate the living cell layers beneath. Almost all bark cracks, forming fissures or plates, because the surface area of the trunk increases as the tree grows larger. In relatively young trees, the bark can stretch to accommodate this expansion, but most bark must eventually crack open. The specific way in which the outer bark layer breaks, and the response the tree produces to deal with the invasion of its tissue, gives us the pleasing array of shapes and textures we see in the woods.

The bark patterns we admire are derived from layers of cells, located just beyond the phloem, called "cork cambium." Their job is to deal with the expansion-induced cracking of the trunk's outer "skin." The bark these layers produce also provides insulation against heat from ground fires and a barrier to prevent invasion by parasitic or pathogenic organisms. At the same time, rough bark surfaces become

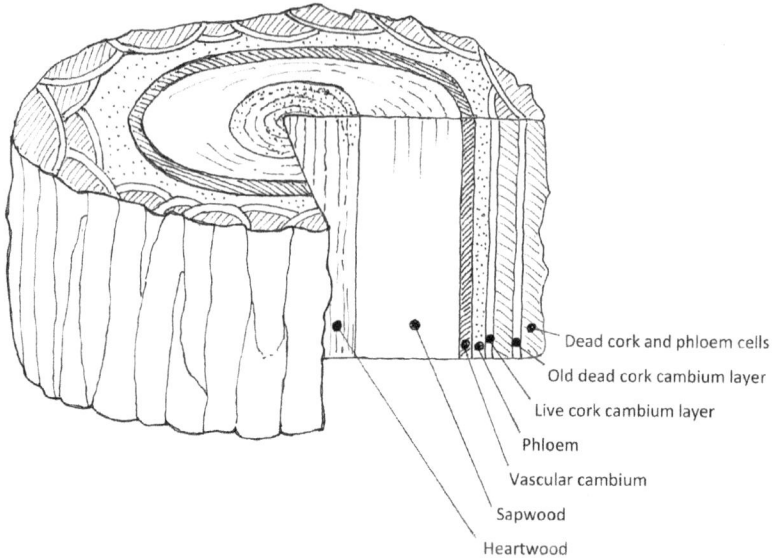

FIGURE 3.8. Cross section of a tree trunk, showing the location of heartwood, sapwood (conductive wood tissue), vascular cambium, phloem, cork cambium layers, and the layer of dead cork cambium and cork cells that forms bark.

the habitat for a wide array of animals and plants. In fact, much of the bark coloration and texture we perceive is derived from the lichens and some mosses that thrive on the bark surface. In most trees, the cork cambium layers arise intermittently as cracks start to develop, rather than following an annual cycle of growth like that of the vascular cambium, xylem, and phloem. The detailed anatomy of cork cambium growth determines the characteristic thickness and roughness of the protective outer bark layer and thus accounts for the great variety of colors and patterns we observe. The way that individual cork cambium layers are spaced and overlap ultimately determines the distinctive appearance of the trees, with deep-seated cork cambium layers producing especially rough and craggier-looking bark, such as we see on our most distinctive oaks and hickories.

Oaks and hickories produce acorns and nuts, which represent a substantial investment in the metabolic resources used to create carbohydrates and fats that, in turn, attract seed predators. Because oaks are such an important component of Ozark forests, and because acorns are so commonly observed along the forest trail, it is worth considering their production as an example of the many subtle factors involved in seed propagation by our principal class of forest tree. In the case of oak seed production, the encouragement of seed predation might appear to be counterproductive as far as reproductive success is concerned. Many rodents and some birds bury nuts in the leaf litter as a means of storage, and some of these creatures will suffer an untimely end and so never return to their seed caches, making up for the many acorns that are consumed. Another part of the reproductive success of oaks and hickories results from the seasonal cycle, with a short-term abundance providing more nuts than a limited population of predators can consume. The collective term for the nut and acorn crop of a forest is "mast." The trees manage to improve the odds of seed survival by a process known as "masting," whereby large mast crops occur only every few years. To some extent, the stress of producing a nut crop in a good mast year programs trees to limit their production in the following year or so. Masting is further enhanced by years in which late freezes damage the tree flowers that appear ahead of leaf-out in the early spring. If there are periodic seasons in which there is no mast crop at all, these events can severely limit the population of seed predators for years to come—leaving enough squirrels to transport seeds to new locations but not so many as to decimate the nut crop. Once you are aware of masting and its role in tree reproduction, you can recognize those autumns when the trail is literally covered by acorns, often with the characteristic shape of one specific red or white oak species. You will also know that this is not a coincidence but an indication of the finely tuned ecological character of mast-producing and nut-predator species that have been interacting with each other for millennia.

A few careful observations of squirrels and jays as seed consumers will show exactly how finely tuned the masting process has become. Some oaks have acorns that take two years to mature, and, in the process, they can be loaded with more lipids but are also full of tannins that create a bitter taste and interfere with seed digestion. Other oaks have acorns that mature in a single season and are loaded with much less tannin, making them the preferred rodent food. The one-year acorns make up for this by sprouting almost before they hit the ground, so that they are exposed to predation for a short period. Squirrels concentrate on consuming the one-year acorns in the fall, then collecting the largely dormant two-year acorns later in the winter. Both squirrels and jays have been observed deliberately killing each seed embryo before storing it, thereby ensuring that the acorn's palatability is not degraded by sprouting. Gray squirrels have even been raised in captivity by researchers to verify that this is instinctive behavior and not something they learn from their parents in the wild.

This brief discussion of the aspects of tree anatomy and seed production that influence our observations in the forest brings us back to the subject of the trees themselves. In the nominally oak–hickory forests of the Ozarks, the oaks are represented by a greater number of species than the hickories. The oaks can be divided into two groups on the basis of a readily distinguishable feature of their leaves. In the "white oak" group the lobes of the leaf are rounded and lack bristles on their tips, while in the "red oak" group each lobe has a bristle (Figure 3.1). The first group includes, in addition to the white oak itself, such other species as post and chinquapin oaks, while the second group includes such common examples as northern red, southern red, black, pin, and blackjack oaks. White and red oaks also differ in the way their acorns mature: white oak acorns ripen in the fall after the flowers are pollinated; whereas red oak acorns take two seasons to mature (Figure 3.9), with the consequence for seed predators described above.

White oak is often the most common oak in Ozark oak–hickory forests and is easy to identify from its leaves, which have five to nine finger-like lobes. This species can grow to be a large tree (more than a hundred feet tall, with a diameter of three feet) on favorable sites. White oak is known to be capable of reaching an age of more than four hundred years, although very old individuals appear to be uncommon in the Ozarks. Post oak, a species that can be recognized on the basis of its distinctly cross-shaped leaves, is the tree most likely to be represented by very old individuals in the Ozarks. On poor sites, where the original forests contained too few trees to be worth logging, it is not unusual to encounter post oaks that were already big trees well before the time of the Civil War. The leaves of chinquapin oak have teeth along their margins instead of lobes. The teeth are pointed but, like all other members of the white oak group, lack any evidence of bristles.

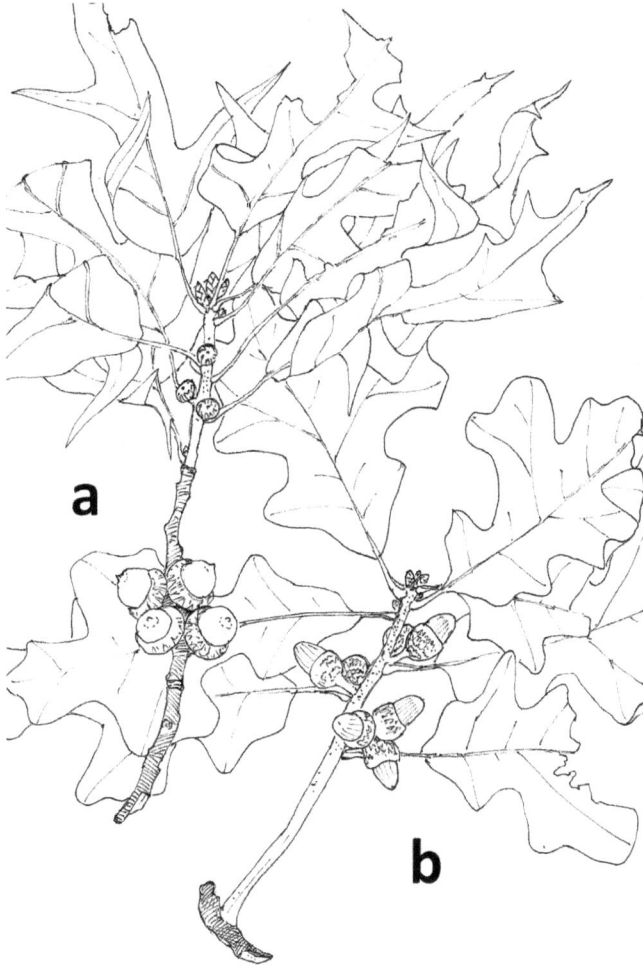

FIGURE 3.9. Examples of (**a**) a species of red oak (southern red oak) with acorns that require two growing seasons to mature (both immature acorns from the current year's growth and mature acorns from the previous year's growth are seen on the tree); and (**b**) a species of white oak (post oak) with acorns that mature in a single growing season.

The gray, flaky bark of chinquapin oak is very similar to that of white oak, but you can often see that the flakes are distinctly rounded, rather than typically rectangular as in white oak. Chinquapin oak is known to prefer soils with a high calcium content and is often associated with limestone; it is rarely seen on soils derived from sandstone. Chinquapin oak also turns out to be our most widespread oak species, ranging all the way from central Mexico to southeastern Canada.

Northern red oaks are among the largest trees found in Ozark forests. The leaves have seven to nine lobes and are relatively large (often five to nine inches long and four to six inches wide). The lobes of the leaf taper gradually from the broad base to the narrowed tip, which has one or more long, bristle-pointed teeth. The leaves sometimes turn a brilliant red in the fall, although in some years the predominant color is brown or yellow-orange. Southern red oaks tend to be smaller than northern red oaks and often occur on poorer sites. The leaves of this oak are rather variable but have a more rounded base than those of northern red oak, and the lower surface is covered with short hairs (the lower surface of the leaf in northern red oak is essentially smooth). In some instances, the leaves of southern red oak are somewhat bell-shaped and have only three lobes, which causes them to be very different from those of northern red oak. The leaves of black oak and northern red oak can be very similar in overall shape, but the former can be distinguished by the presence of tufts of rusty brown hairs in the forks of the veins on the lower surface of the leaf. While many Ozark oaks can be found in a variety of habitats, most have some preferences, and a few, such as blackjack oak and pin oak, are generally restricted to specific kinds of sites (Table 3.1).

Several different types of hickory are found in Ozark forests (Figure 3.2). The most common of these are shagbark hickory, mockernut hickory, and black hickory. Shagbark and mockernut hickories are both large trees (reaching a hundred feet on favorable sites) when fully mature, while black hickory is smaller and rarely exceeds seventy feet. The single most distinctive feature of shagbark hickory (and the one that accounts for its common name) is how its outer bark separates from the trunk but remains attached as a series of long strips. Shagbark hickory is a characteristic tree of deep, rich soils and is often found on sheltered benches beneath steep bluffs or on terraces along smaller mountain streams. In contrast to that of shagbark, the bark of the other Ozark hickories is fissured. The fissures vary from sinuous and barely noticed on bitternut hickory; to moderately ridged in a kind of braided pattern on mockernut hickory; to greatly exaggerated, with rough rectangular plates intersecting in a diagonal pattern, on black hickory (Figure 3.2). The lower surface of the leaf of mockernut hickory is covered with dense short hairs, which is usually enough to identify this species. In addition, the leaves have a strong, resinous odor when crushed. Black hickory tends to be found on poorer sites than either shagbark or mockernut hickory, the leaflets are more narrow, and the fruit is pear-shaped and has a thin husk. For both shagbark and mockernut hickory, the fruit is more or less globose and has a thick husk. Bitternut hickory is unlike our other hickories in preferring rich, moist soils of bottomlands and at the immediate base of sandstone ledges. Pignut hickory ranges throughout the Ozarks on dry upland soils but is much less common than the other hickories. All of our native hickory species bear nuts that are technically edible, but only shagbark bears nuts with thin enough shells and large enough kernels to make them worth developing into commercial nut-crop varieties that you can order from a nursery catalog. Resourceful Native Americans learned to utilize hickory nuts by immersing the cracked shells in boiling water so that the fat-rich oil from the nut meat could be collected where it rose to the top of the boiling mass. When it comes to hickory identification, there can be some variation in leaf shape within hickory species, but each has a predominant number of leaflets, varying from five to nine, on each leaf (Figure 3.2). That observation and a knowledge of the characteristic bark patterns are usually enough for identification in the field.

Few of the other trees found in Ozark forests have compound leaves, and these tend to be relatively easy to identify (Figure 3.3). Black walnut is typically confined to sites with relatively rich and moist (mesic) soils. Although walnuts are members of the same family (Juglandaceae) as the hickories, the two types of trees are easily distinguished by the numbers of leaflets that make up their compound leaves. The compound leaves of walnut have fifteen to twenty-three leaflets, which is higher than the number found in any hickory. Indeed, for the more common species of hickory, the number of leaflets is usually no more than nine. Moreover, in walnut, the largest leaflets are located toward the center of the leaf, while in hickory the terminal leaflet is usually the largest. The fruit produced by the walnut is a relatively large nut that has a brownish-green, semi-fleshy husk. Be careful when handling the green husks of freshly fallen black walnuts, because they contain a sap that will stain your hands black. The kernel of the nut is highly edible, and the wood of walnut is exceedingly valuable for its use in making furniture.

White ash is similar to walnut and hickory in having compound leaves, but these leaves differ in one important respect: the leaf arrangement in white ash is opposite, rather than the alternate condition characteristic of both walnut and hickory. The combination of opposite compound leaves is not common in trees, and white ash shares this feature with only one other tree found in the Ozarks, the box elder (which is actually a maple). However, the two trees differ in a number of other ways, including the shape of the leaflets (distinctly toothed in box elder but almost entire in white ash). White ash gets its name from the pale lower surface of the leaflets, which are lighter in color than the upper surface. Green ash, which also occurs in the Ozarks, is very similar in appearance to white ash, but the upper and lower surfaces of each leaflet are similar in color. The two types of ash also tend to be found on different types of sites, with white ash commonly occurring in well-drained forests throughout the Ozarks, while green ash is largely restricted to riparian forests and areas that have been subject to disturbance.

Sugar maple and red maple are common and widespread trees in the Ozarks. As already noted, maples have opposite leaves with palmate venation, and this combination of features makes them easy to recognize. The leaves of both types of maple have three (more rarely five) distinct lobes, but the margin of the leaf of red maple is coarsely toothed, whereas the margin of sugar maple has few if any teeth. Moreover, the bottom of the depression between any two lobes (referred to as a "sinus") is V-shaped in red maple and U-shaped in sugar maple. The fruit produced by both types of maple is a distinctive, two-seeded, winged samara. Red maple is most often a relatively small subcanopy tree in the Ozarks that only occasionally becomes a large canopy-dominant tree, though it can be a major forest tree farther north. The taxonomy of the sugar maple is in flux because the tree exhibits a significant amount of variability throughout its range. Some authorities want to make a separate species out of the eastern Oklahoma variety, while the shrubby sugar-maple-like trees found in western Oklahoma are sometimes equated with the bigtooth maple of Utah and west Texas. Ozark sugar maples might just be a giant hybrid swarm.

The three other deciduous trees regularly present in Ozark oak–hickory forests are black gum, elms, and black cherry (Figure 3.3). Black gum is nearly ubiquitous on all sites and sometimes grows as large as any other tree in the forest. Because black gum wood is not valuable, large specimens can often be found in second-growth forests otherwise regenerating from past logging. Tree-ring studies show that black gums can be more than four hundred years old; among Ozark trees, only ancient post oaks are consistently older. Old black gums develop a distinctive "alligator skin" bark and are most often hollow, providing homes for raccoons and other wildlife. Black gum fruits are attractive, bluish berries that can sometimes be seen from the vantage of a rock ledge, where you can look directly into the crowns of trees growing at the base of the cliff. The leaves of some black gums take on an especially brilliant scarlet in fall, often before other trees have even begun to turn color.

There are three species of elm in the Ozarks: American, slippery, and winged. These trees can be difficult to tell apart, and most hikers will simply recognize them as elms. Although somewhat smaller than the other two, winged elm is the most consistently abundant of the three in upland oak–hickory forests. Winged elm leaves are variable in size but are usually smaller than the leaves of other elms, and winged elm twigs often (but not always) have distinctive corky "wings" lining them (Figure 3.10). All of our elms have small, rounded, one-seeded fruits called "samaras."

Black cherry is a tree that is rarely common in the oak–hickory forest but is consistently present throughout. This is another wide-spread species, ranging from southern Mexico to Quebec, and it has some of the most valuable wood of any tree, providing an attractive reddish color and wavy grain for fine furniture and woodwork. The tree is especially conspicuous during excursions in the Ozarks, since it is easily recognizable by its black, scaly bark.

All of the species mentioned thus far are broadleaf trees and belong to the major taxonomic assemblage of plants known as "angio-sperms" (the flowering plants). However, a few species belong to an

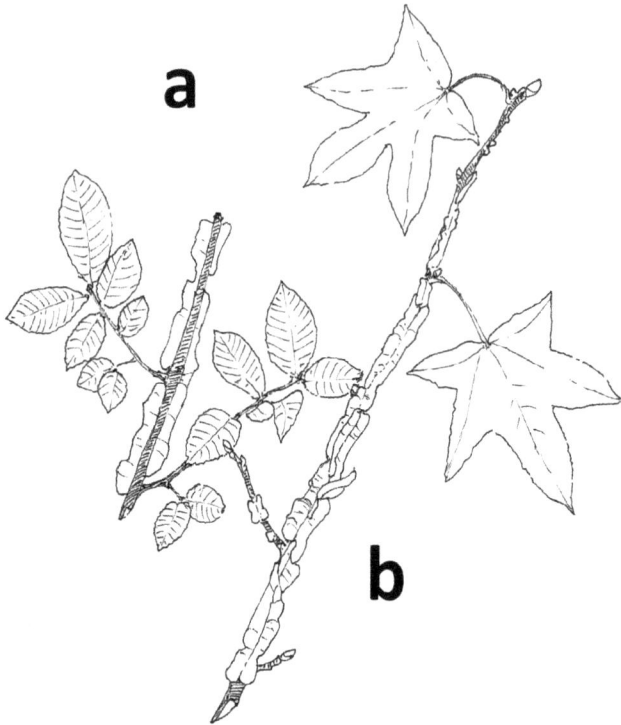

FIGURE 3.10. Examples of two common Ozark trees in which branch tips often develop corky ridges or "wings": (**a**) winged elm and (**b**) sweetgum.

entirely different taxonomic assemblage, the "gymnosperms," a much more ancient group of plants that produce cones instead of flowers. In Arkansas, the most common and widespread gymnosperms are red cedar, various species of pine, and bald cypress. The only naturally occurring pine in the Ozarks is shortleaf pine, a fast-growing species that can exceed a height of one hundred feet on favorable sites (Figure 3.11). Like all pines, this tree is easily recognized from its needle-like leaves. These are slender, yellowish-green in color, three to five inches in length, and occur in clusters of two (or sometimes three) on the twigs. The seed cones of shortleaf pine are about two inches in length, and each of the individual scales that collectively make up the cone has a small, sharp prickle. This pine is a major source of wood pulp and lumber in the Ozarks and in other parts of its range. Shortleaf pine is most often seen as second growth on land where valuable old-growth pine timber was once harvested (Figure 3.11, c). Native shortleaf pine can be distinguished from loblolly pine, which is widely planted as a commercial forestry tree in the southern Ozarks, because loblolly consistently has clusters of three needles and significantly larger cones. Occasional old pines can be seen in locations where they were inaccessible to loggers or where large individual trees were left because they were not suitable for timber (Figure 3.11, a, b). Shortleaf pine is unusual in that a damaged tree can regenerate branches after a windstorm: long-dormant buds buried in the bark can sprout to produce new twigs (Figure 3.11, d). By contrast, most other pines have no way to generate a new crown after severe weather has stripped away the branches from a battered trunk. Bald cypress is only marginally present in Ozark uplands and is hardly relevant in this discussion. Red cedar is much more a part of the Ozarks and is both widespread and abundant—at least at present. However, the cedar is an important subject in relation to prehistory and forest response to historical disturbance. It is rarely present in older forest growth and was originally restricted to special habitats such as exposed bluffs and rock ledges. Red cedar will figure prominently in some of our subsequent chapters.

Ending this survey of Ozark forest trees, we come back to the basic question of why there are so many different species of oak and hickory. The average hiker struggles with learning to tell the various oaks from each other. This raises questions about exactly how they remain separate species when they live together so closely, shedding pollen at the same time among so many closely related species. Modern DNA analysis provides a way to look into this. Geneticists have collected samples from sets of oak trees nominally representing three different species. Leaves and acorns were classified according to the noted characteristics of each species, and many showed signs of overlap. Surprisingly, very little genetic interbreeding was apparent from the DNA analysis. Leaves and acorns may vary from the standard model, but the oaks apparently know exactly who they are.

That brings us to one of the truly great mysteries concerning forests in general, and Ozark forests in particular. That is the subject of diversity. The Appalachian region is noted for its great botanical diversity, much of which extends into the Ozark region. The exact nature

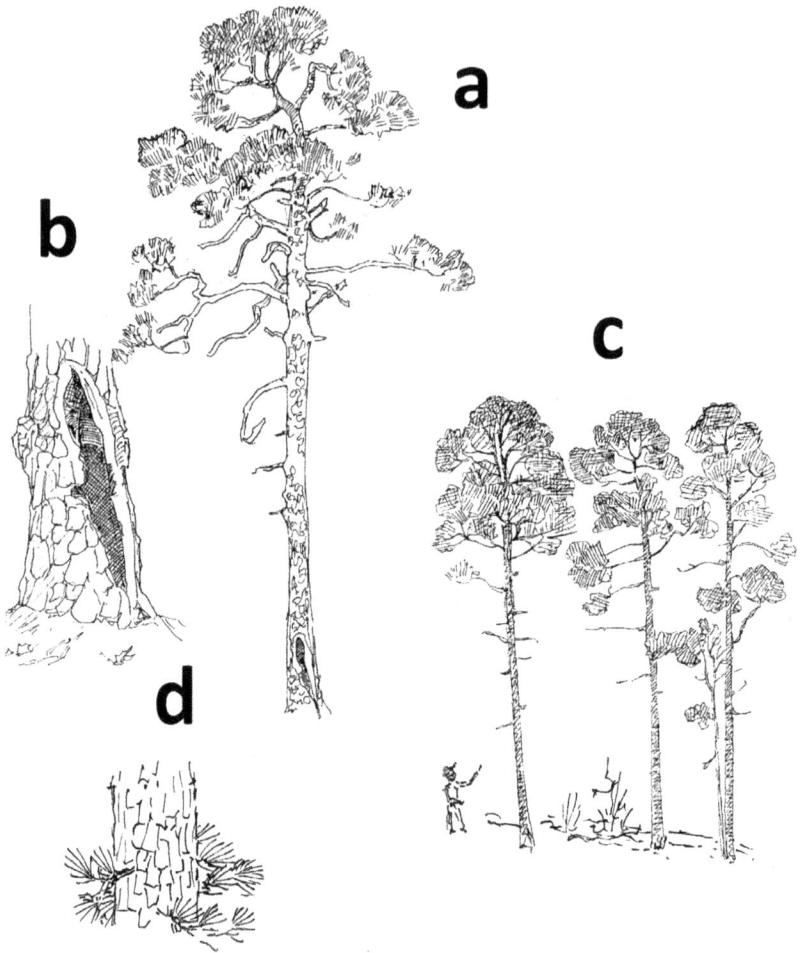

FIGURE 3.11. Shortleaf pine in the Ozarks: (**a**) an old-growth tree; (**b**) detail of
fire scar from repeated burnings at the base of the same tree; (**c**) typical stature of
second-growth individuals in Ozark forests today; and (**d**) detail illustrating this
species' unusual (for a conifer) ability to generate new branches from the bare trunk
of a damaged tree.

of the forces that drive diversity remains a subject of serious scientific debate. Many years ago, it was simply understood that trees and
other plants were adapted to fit specific habitat niches. Thus, tropical
rainforests were thought to have become centers of extreme diversity by virtue of having been unchanged for eons, allowing plants to

proliferate into many different specialized species. One famed North American forest ecology expert, Lucy Braun, used the diversity of the flora in southeastern America as evidence in arguing that the climate south of the glacial boundary had been virtually unaffected by the coming and going of ice sheets. That would have provided the long era of climate stability required to generate our forest diversity. We now know that exactly the opposite is true—glacial periods wrought great climate changes at all latitudes. Colder eras were accompanied by drier climate, such that even tropical rainforests contracted significantly. In fact, one can argue that the changing climate worked to drive diversity by isolating trees into local populations where they developed their own special genetic traits as they waited out those hostile dry climates. But even here there is a problem. It turns out that in the most diverse tropical forests with several hundred different tree species, most of those species turn out to be generalists and are not tied to any specific niche at all. As in the case of so many other scientific questions, the devil lies in the details. Science has to unravel eons over which climate has been shifting back and forth between extremes. Tree species have been forced to migrate, trees have been subject to inbreeding and mutation while isolated in multiple local refuges, and all the while trees are evolving elaborate biochemical defenses to deal with an ever-advancing army of pathogens. The Ozark forests we see on casual outings are the result of millions of years of intricate interactions among fiercely competing species and their biological enemies.

CHAPTER 4

Other Forest Realms

Although the forests of the Ozarks are broadly classified as oak-hickory, there are several habitats within the region that have their own distinctive varieties of tree species and other plants, growing in associations that look very different from the typical scene depicted in Figure 1.1. Several of these habitats, such as deep ravines and open glades, are generally considered interesting scenic destinations for outdoor enthusiasts. Other habitats, such as roadsides, old fields, and river bottoms, are just parts of the landscape that hikers encounter on almost any excursion. All of these habitats represent relatively large-scale associations of trees within the Ozark upland forest mosaic. But there are other, much smaller habitats (sometimes called "micro-habitats") that are also of considerable interest. Prominent examples include the intricate rock gardens of mosses, ferns, and specially adapted plants that can be found on the surface of sandstone blocks tumbled into ravines, or the mounds of exposed mineral soil raised up by the roots of overturned trees that serve as preferred germination sites for some plant species. The general discussion of the oak-hickory forest association given in Chapter 3 needs to be augmented by a closer look at some of the other interesting habitat realms—both large and small—embedded in the otherwise oak-dominated forests of our region.

The geological framework of the Ozarks as an uplifted plateau composed of erosion-resistant sandstone and limestone beds naturally leads to the generation of deep ravines and canyons with nearly vertical rock walls, formed where the hardest sedimentary beds crop out in valley walls (Figure 4.1). These cliffs and ledges form the most visually prominent parts of the habitat, but the actual erosive process is controlled by the presence of large rock blocks that fill the drainage channel of the steeply dropping stream within ravines. The hard rocks serve to armor the streambed against erosion, ensuring that the head of the ravine remains extremely steep. This is a natural consequence of the way that the cliffs weather, with blocks of hard rock separating along rectangular, stress-induced joints under the relentless grip of gravity. You can actually witness the process on a typical hike if you go out on the first warm days after a long period of subfreezing tempera-tures. The freeze–thaw cycle works boulders loose, and you can often hear them crashing against other boulders and splintering trees as they tumble down. On calm, warm, late-winter days, this sound carries a considerable distance up and down ravines. Even if the probability of being harmed by such rare rock-fall events is exceedingly low, the sound carries so far that you have a good chance of hearing it happen on warm winter days if you time your excursion properly.

The most prominent characteristic of Ozark mountain ravines is the shielding of the environment from direct sunlight and the con-stant presence of moisture seeping from the ravine walls. The ravine environment is so completely different from that of the upland forest that tree reproduction depends on factors specific to this environ-ment and results in a largely different forest composition from that of the oak–hickory association. One oak species, the northern red oak, is relatively well adapted to moist and shady environments, and it is the one oak commonly found in ravines. Black gum is also common in ravine habitats, and bitternut hickory is sometimes found at the base of rock ledges that outcrop high on ravine walls. Three other tree species—beech, basswood, and cucumber tree—otherwise dominate ravine environments (Figure 4.2). American beech reaches the north-western limit of its range on the southern and eastern fringes of the Ozarks, but it can often be the single most common tree in the ravine environment in that part of our region. Beech trees in our forests are very susceptible to heart rot decay (see Chapter 13), and large beech trees are almost always hollow. Beech is also unusual in that its roots can generate new stems at a considerable distance from the base of the tree, whereas many other trees can generate new sprout stems only from the base of the main trunk. This may help account for the way in which beech can locally dominate ravine habitats.

Basswood has a much more northern distribution than beech and

FIGURE 4.1. Typical mesic Ozark ravine habitat with steep slopes, sandstone blocks armoring the streambeds, and a deeply shaded environment.

is often considered a tree that replaces beech where northern hardwood forests extend beyond the latter's range in Minnesota. Basswood has relatively soft and weak wood, which makes it ideal for wood carvers but results in frequent storm damage. Basswood compensates for this by readily generating new stems from the base of the original tree. You can often see smaller secondary trunks growing at the base of

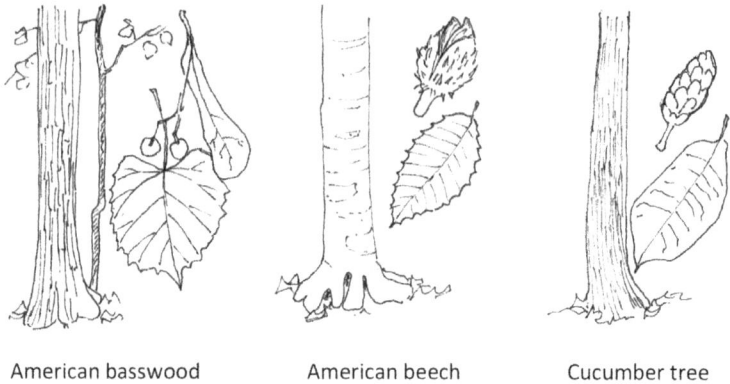

American basswood   American beech   Cucumber tree

FIGURE 4.2. Leaves, seeds, and bark of three common tree species in the Ozarks that
are restricted almost entirely to ravine habitats. Note the interesting basal apron
of roots typical of large beech trees, and that most basswood trees have one or two
smaller stems arising from the base of the tree.

large and otherwise healthy basswood trees. Even more distinctive is
the ring of several basswood trees that can sometimes be seen growing
around the now empty space where a large basswood tree once grew
(Figure 4.3).

Cucumber tree, a species of magnolia, is the Rodney Dangerfield
of trees in Ozark mountain ravines—it usually gets no respect.
Although sometimes the single most common tree in our secluded
ravines, the cucumber tree (named for its dark green, knobby fruit
that looks something like a deformed cucumber) hardly seems to be
noticed by hikers. It doesn't grow especially large and has a generally
nondescript, medium brown, broadly ridged bark. It has relatively
small, greenish-white flowers that are found high in the crown of the
tree and thus fail to provide any kind of floral display. Most hikers
think of magnolias as smaller trees with attractive flowers, so this particular magnolia regularly escapes notice.

Although the dark and moist ravine environment is itself an interesting world to explore, the boulders within the ravine provide a very
special microhabitat (Figure 4.4). The generally moss- and lichen-
covered surfaces of deeply shaded sandstone boulders have a unique
association of plants and appear as exquisitely beautiful little rock

FIGURE 4.3. Clump of basswood trees growing around the hollow where a large basswood once stood before the present ring of trunks was generated by basal sprouting from the original tree.

gardens. Some of the plants that are regularly found in these moss gardens are eastern columbine, wild hydrangea, alum root, various species of violets, and ferns such as polypody and walking fern (ferns are discussed further in Chapter 11). Some trees that produce relatively small seeds requiring bare soil for germination, such as elm and sweetgum, can establish numerous seedlings in moist mosses on rocks near streams. Almost all of these seedlings are doomed because their roots cannot find a way into the soil, but occasionally small trees can persist where you see their roots draped over the sides of rocks in an interesting configuration. One small tree that is especially suited for getting by in this environment is the red buckeye, which commonly inhabits moist ravine environments in the Ozarks. Hikers venturing into secluded ravines enjoy the pristine environment of tumbled boulders and rock ledges, but they should also take the time to look at the intricate little plant communities situated on the individual mossy boulders.

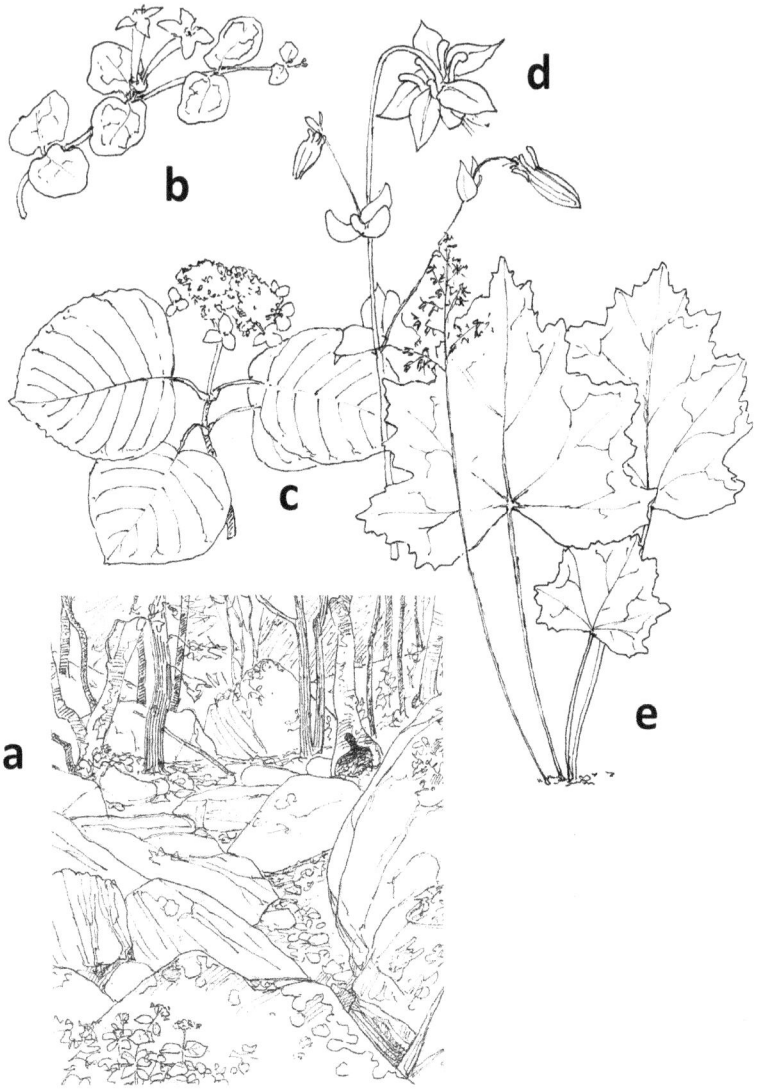

FIGURE 4.4. Ravine rock-garden setting (**a**), where interesting plants include partridgeberry with its fragrant, pale pink pairs of flowers (**b**); wild hydrangea (**c**); eastern columbine (**d**); and alum root with its drab, pale flowers offset by intricately shaped leaves that often have an attractive dark red or purple color (**e**).

The one other Ozark habitat that we often hear about is the glade (Figure 4.5). Although the term "glade" is often used, most outdoor enthusiasts are hard pressed to provide an exact definition. In his book *The Terrestrial Natural Communities of Missouri*, Paul Nelson defines glades as "open, exposed bedrock areas dominated by drought-adapted herbs and grasses in an otherwise woodland or forest matrix." Other naturalists define a glade as any local environment within a for-ested ecosystem where the soil substrate is extremely thin or otherwise inhospitable to trees. In that sense, there can be shale glades (Figure 4.6) where something like a soil exists, but it is deficient in nutrients and in the moisture retention capacity that can otherwise support a closed-canopy forest composed of the trees typically found on Ozark uplands. Glades are divided into subcategories based on the mineral substrate on which they occur: limestone, dolomite, sandstone, chert, shale, and igneous (nonsedimentary basement). For typical hikers, the most important distinction to make is between those substrates char-acterized by acidic (sandstone, shale, and granite) or basic (limestone and dolomite) chemistry. Red cedar is the most prominent tree found in the glade habitat, and its expansion in both that habitat and the sur-rounding forest, as a result of human manipulation of the landscape, will be an important subject in later chapters.

Reports from early European visitors to the Ozarks (see Chapter 5) make it clear that glades and "barrens" were then an important part of the landscape. Ecologists now agree that fire was required to maintain glade communities, and that the exclusion of fire after European settlement has caused a significant degradation of glade environments. With the passage of time, accumulation of organic debris has allowed enough soil to develop on the fringes of glades to allow forest trees and shrubs such as red cedar, oak, and winged sumac to greatly reduce the amount of exposed bedrock in the glade habi-tat. Thus, the glades we can visit today are greatly changed from the glades that the first European visitors saw. With the generally domed nature of the Ozark uplift, glades arise where harder, erosion-resistant layers intersect the surface. For this reason, the glades of the Boston Mountains are primarily composed of sandstone, whereas limestone and dolomite glades are abundant on the Springfield and Salem pla-teaus, and igneous glades are restricted to the St. Francois Mountains where underlying basement bedrock is exposed.

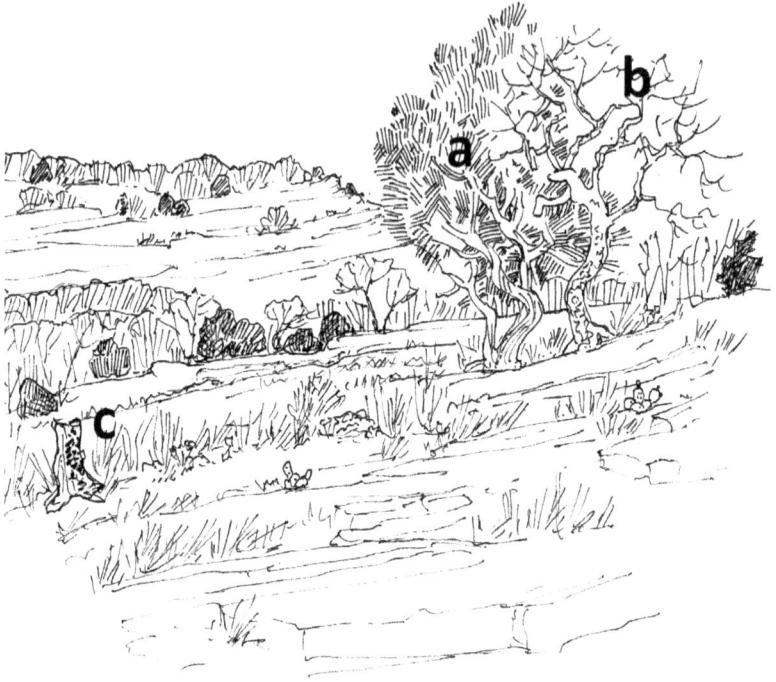

FIGURE 4.5. View of a cedar glade, showing exposed rock ledge and glade habitat on the Salem Plateau ridges in the background; the two species of trees most common on limestone (basic) glades are red cedar (**a**) and post oak (**b**). The deeply charred wood on a long-dead oak stump (**c**) shows that fire has been required to maintain the open glade condition, and the relatively low density of red cedar around the edges of this glade indicates that fire has been frequent enough in the past to keep that invasive tree at bay. Today, most such glades are densely overgrown with red cedar unless labor-intensive removal projects have been completed.

The exposed bedrock of glades provides an interesting landscape for the Ozark hiker to survey (Figure 4.5). Ecologists are interested in the fact that glades provide isolated "islands" of habitat for species specially adapted to that environment. Some of these species were probably left behind when significant open grasslands in the Ozarks were part of a greatly expanded prairie environment, during a period about six thousand years ago when the midwestern American climate was somewhat warmer and drier than it is today (see Figure 2.17B). Other species have probably developed over time to suit this specific environment. Red cedar was once largely restricted to the edges of

FIGURE 4.6. Shale-glade habitat in winter; the glade has an open stand of very scrubby blackjack oaks, with a thin, discontinuous layer of leaf litter over rock shards and small shale outcrops.

glades before greatly expanding its distribution after settlement. Acidic glades frequently have post oak and blackjack oak, farkleberry, and fragrant sumac in addition to red cedar. Basic glades have chinquapin oak, dwarf hackberry, winged elm, and Shumard oak. Most important for outdoor enthusiasts is the dwarfed and contorted shape that these trees take when forced to grow in the severe glade environment.

Glades are home to a number of interesting wildflowers and other low-growing plants (Figure 4.7). These include blazing star, verbena, orange puccoon, Indian paintbrush, and eastern prickly pear. A number of interesting animals and birds are found in glade habitats. The roadrunner, familiar in our suburban environment with its man-made openings, was once restricted to glade openings. The colorful painted bunting is also found in the open habitat provided by glades. Collared lizards and several snake species are glade inhabitants, too. Hikers venturing out onto trails that pass over and through scenic glades should keep their eyes open for these special local inhabitants as a way of enhancing their outdoor experience.

River bottoms provide yet another habitat that is distinctly different from the oak- dominated upland. One of our oaks, the pin oak, is specifically adapted to growing in bottomland soils and can be found on river bottoms along with Shumard oak. The latter has a kind of split personality similar to that of southern red oak, with one race adapted to dry rocky outcrops and another that prefers rich river-bottom soils. Bitternut hickory also seems to do well along the course of smaller headwater creeks. However, four trees are most often associated with the banks of Ozark mountain streams: sycamore, sweetgum, river birch, and box elder (Figure 4.8). All four have relatively small seeds that are widely distributed by wind. Such seeds have difficulty penetrating thick leaf litter, and these species do not grow well in shaded environments. Rivers erode laterally as they meander (Chapter 8), leaving exposed gravel bars behind as they move across their valley bottoms. Widely disseminated seeds can quickly find these newly exposed patches of soil and become established. These trees can then grow rapidly to shade out the few oaks or hickories that might find their way into the seedling mix. One often sees a mosaic of river-bottom stands composed of patches of a single tree species such as box elder or sweetgum, where one propitiously located seed-source tree has managed to generate a large crop of offspring on a newly exposed section of river sediments.

Another common habitat within the oak–hickory forest is the roadside strip and adjacent abandoned fields (Figure 4.9). These are not the natural or pristine environments that we usually want to experience on our outdoor adventures, but they are unavoidably part of the hiking experience. Red cedar is a ubiquitous component of this habitat (Chapter 10) as a result of its prolific production of seeds ideal for dissemination (fertilizer and all) by birds and the relative unpalatability of its foliage for livestock and deer. White ash also often seeds into abandoned pastures in nearly pure stands. Three other tree spe-

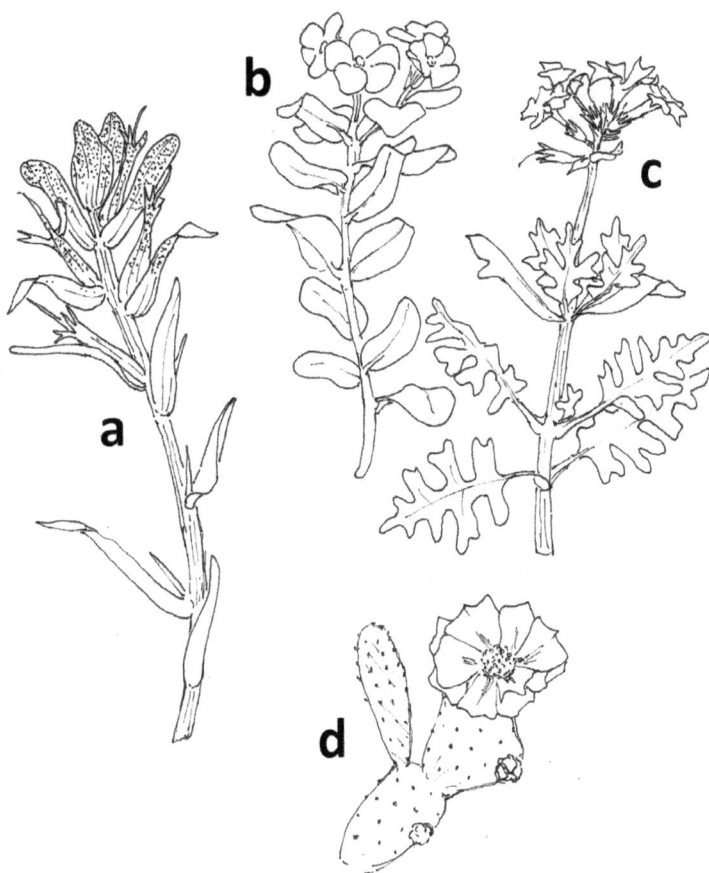

FIGURE 4.7. Some of the most common species of wildflowers found on rock ledges and in cedar glades: (a) Indian paintbrush, (b) orange puccoon, (c) verbena, and (d) prickly pear.

cies stand out as most often seen in the disturbed areas around highways: black locust, honey locust, and hackberry (Figure 4.10). Both locust species have spread far beyond their original range because they are so well adapted for invasion of disturbed areas. Black locust has distinctive ridged bark and is relatively short lived, such that dead or decadent black locust trunks may be mixed in with other trees that are still reaching maturity. Honey locust usually comes equipped with swarms of very vicious-looking thorns studding the trunk and branches (Figure 4.11, a). Hackberry has interesting and unusual bark

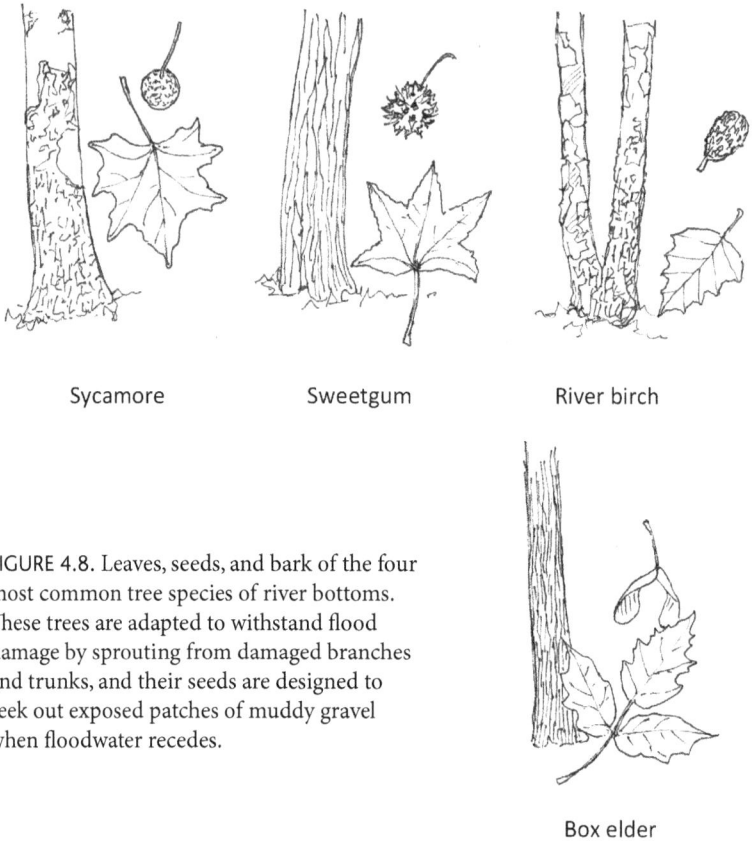

Sycamore   Sweetgum   River birch

FIGURE 4.8. Leaves, seeds, and bark of the four most common tree species of river bottoms. These trees are adapted to withstand flood damage by sprouting from damaged branches and trunks, and their seeds are designed to seek out exposed patches of muddy gravel when floodwater recedes.

Box elder

before it reaches maturity, with vertical rows of distinctive wart-like protrusions (Figure 4.11, b), but at full maturity the dark, scaly bark is not too different from that of black cherry. Hackberry lives long enough to become a forest giant if given the opportunity (Figure 4.12).

A minor forest microhabitat worth examining is the "tip-up mound" produced when wind topples a mature tree (Chapter 7). The roots erupt to produce a mound of exposed mineral soil carried with the overturned root plate and create an adjacent hollow where the roots once were situated (Figure 4.13). The bare soil provides a seedbed that is quite different from the thick layer of leaf litter and organic material that characterizes the surrounding forest. Hikers can take a moment to look at the tree seedlings and herbs that use this

FIGURE 4.9. Typical roadside view of an adjacent abandoned field in the Ozark uplands in early autumn. Common trees of the disturbed roadside habitat are black locust (**a**), hackberry (**b**), and black cherry (**c**). White ash and red cedar (**d**) are seen invading an adjacent abandoned field. Roadside areas are often invaded by alien species such as the ailanthus (**e**) and a dense understory thicket of privet and Japanese bush honeysuckle (**f**). The open areas and ditches immediately beside the pavement often have blackberry brambles (**g**), ox-eye daisies (**h**), cattails (**i**), and joe-pye weed (**j**). The roadrunner in the foreground is an open-area species that has adapted well to our modern roadside habitat.

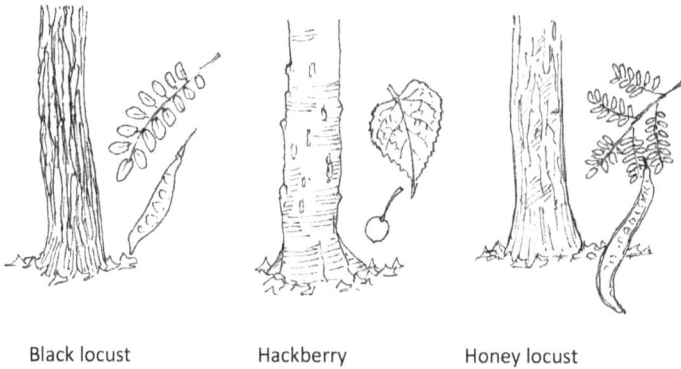

Black locust   Hackberry   Honey locust

FIGURE 4.10. Leaves, fruits, and bark of three common species of roadside trees in the Ozarks. Black locust and honey locust once had much more limited distributions but are now widespread because so much roadside and abandoned-field habitat has become available. Hackberry was once primarily a river-bottom tree but has likewise found itself well adapted for invasion of roadsides and fields.

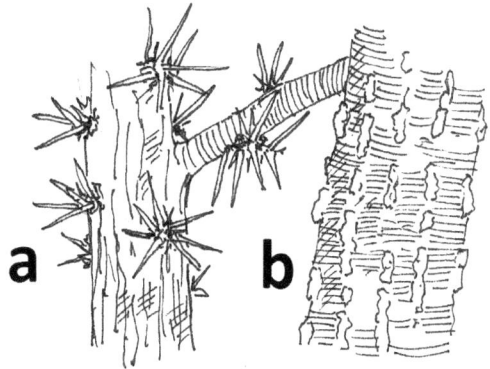

FIGURE 4.11. Distinctive features of two roadside trees: (a) dense clusters of large thorns on the trunk and branches of honey locust and (b) pronounced warty lumps on the otherwise smooth bark of a young hackberry.

microhabitat. It is also common for various wildlife species such as foxes, coyotes, and bobcats to develop dens by burrowing underneath the overturned root plate. The significance of exposed mineral soil in establishing specific tree species can be seen on the trail, where the walkway itself produces a minor strip of bare mineral soil, and trees that are not typically part of the mature upland oak forest can become established (Figure 4.14). A dramatic example of this is provided by the alley of stately sweetgums that lines the Ozark Highlands Trail where it follows an old railroad grade near the village of Cass in northwestern Arkansas.

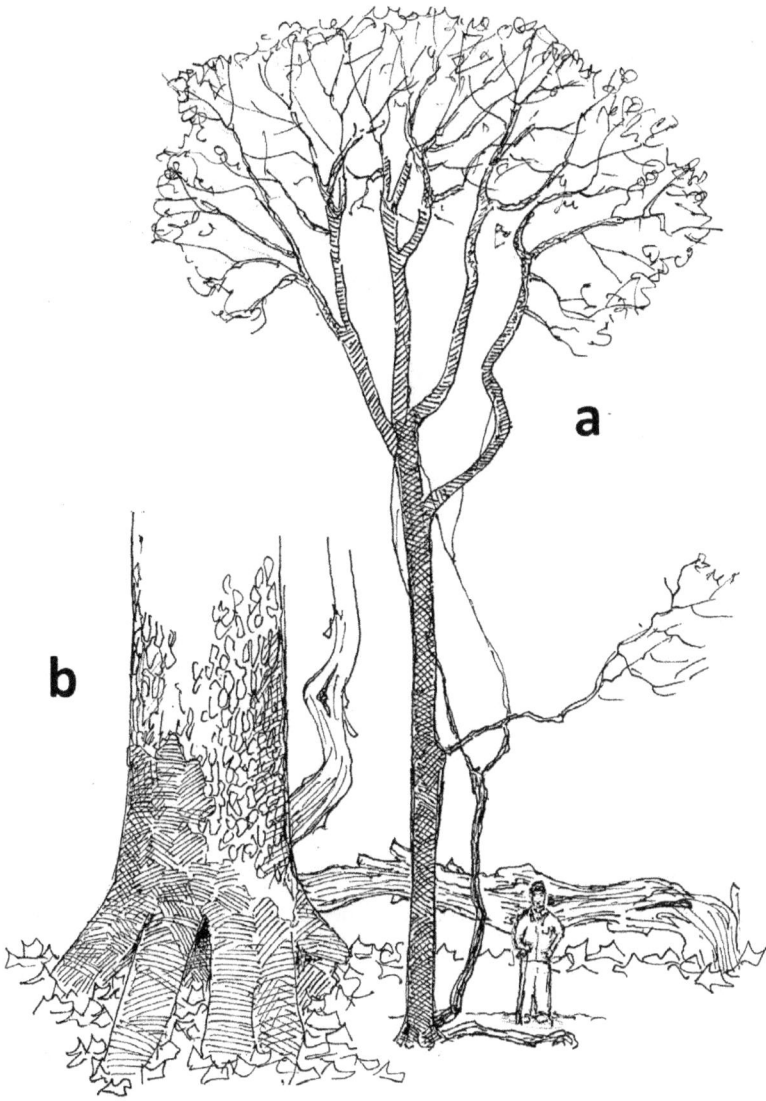

FIGURE 4.12. A majestic hackberry in full maturity, on Lee Creek in Devil's Den State Park, Arkansas: (**a**) profile (with hiker for scale) and (**b**) detail of the base of trunk.

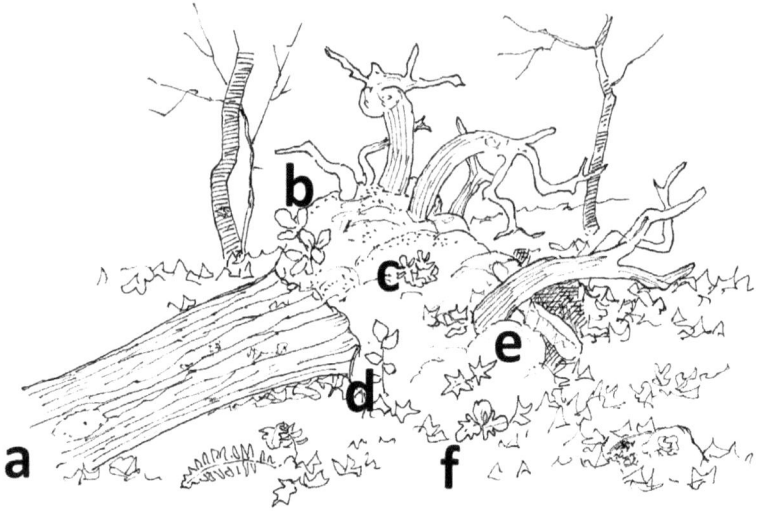

FIGURE 4.13. "Tip-up mound" microhabitat with exposed mineral soil, allowing the establishment of tree seedlings and herbs that could not otherwise seed into the thick leaf litter of the forest floor. Here, a red oak (**a**) toppled in a windstorm has thrown up a mound of mineral soil where horsemint (**c**) and three-lobed violet (**f**) will be flowering, and where seedlings of blackberry (**b**), serviceberry (**d**), and sweetgum (**e**) have become established.

One other interesting Ozark mountain microhabitat is the seepage area where groundwater flow provides abundant moisture throughout the driest parts of summer. Shaded ledges can have water trickling over them where it flows out of bedding planes in the sandstone or limestone outcrop. There can be lush growth of mosses and ferns, such as the delicate maidenhair fern. This is the preferred habitat of the sharp-wing monkeyflower, cardinal flower, and jewelweed. The latter grows with exuberant abundance in this environment. Hikers can amaze children who accompany them on hikes near seepage areas by showing how the half-inch-long, bean-like jewelweed pods literally explode when touched, releasing the tensed-up springs that have developed inside the ripe pod and thereby ejecting the seeds (see Figure 11.5, d). The rare yellow monkeyflower can also be found around seepage areas in the Ozarks, far from its more typical habitat in the Rocky Mountains. Perhaps the most spectacular flower to be found here is the tall, late-blooming great blue lobelia (Figure 4.15).

FIGURE 4.14. Sweetgum saplings along a forest trail. Although seedlings can start growing in the exposed mineral soil on the edge of a trail, they generally never become established in the deep leaf litter of the open forest.

The moist habitat allows this unexpectedly vibrant flower to bloom in late summer, when almost all other floral displays are long past and this splash of bright color comes as a pleasant surprise.

Even though the Ozark forest can be defined as mostly a mosaic of oak, hickory, and pine, with a few other tree species mixed in, the forest we hike through has a lot of local variation that can be observed along the way. You can add to your insight and enjoyment by taking the time to note how differences in local environment—whether at the landscape scale or immediately underfoot—illustrate the environmental forces at work in determining the nature of Ozark woodlands.

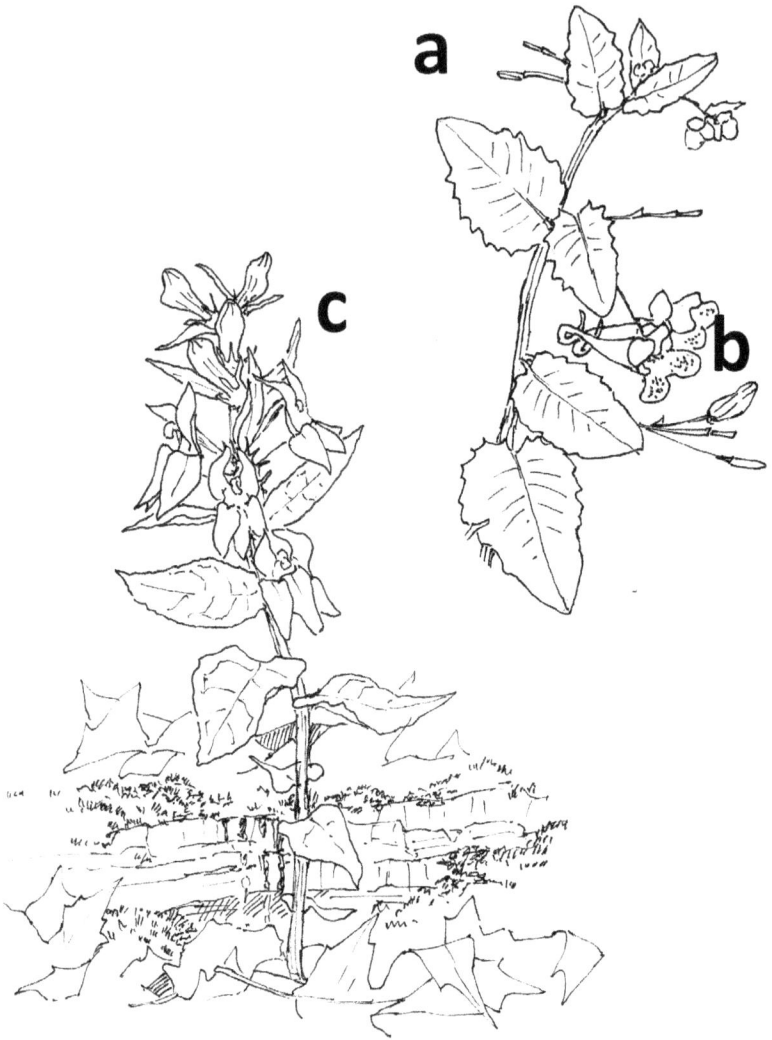

FIGURE 4.15. Jewelweed (**a**) is far and away the most common flower growing where water seeps out at the base of rock ledges along streams and small rivers. The oblong seed pods have an interesting spring-loaded way of ejecting seeds (**b**). However, the giant blue lobelia (**c**) is the most spectacular of the late-summer flowers found in this habitat. The lobelia shown here is the relatively short and stunted form found in deep shade under dense forest cover. Plants growing in full sunshine can have stalks nearly three feet tall holding dozens of flowers, so it should be no surprise that garden cultivars have been developed from this particular species.

CHAPTER 5

The Original Forests of the Ozarks

Many outdoor enthusiasts seek the most pristine and unspoiled locations for their hiking excursions. When visiting such places, it is natural to wonder just how representative these places are of the original, virgin forest found by the first European visitors to the Ozarks. The term "virgin forest" has connotations of great swaths of ancient trees, but we know that Native Americans had already been altering their forest environment before Europeans arrived on the scene, and even areas free from human interference were subject to other disturbances. Reconstruction of our primeval Ozark forest environment is thus a bit more complex than one might think. Still, it is an interesting exercise for hikers to consider how closely the most pristine-looking Ozark locations we can visit today resemble their "unspoiled" counterparts from a few centuries ago.

We have already considered the presettlement forest in our discussion of past climate changes in our region in Chapter 2. The forest reconstructions in Figure 2.17 were based on insights from studies of pollen and macrofossils. Pollen proportions from the uppermost late-prehistoric deposition layers at locations such as Cupola Pond indicate a forest dominated by oak, with considerable pine and hickory, along with a bit of grass pollen. The one major problem with the use of pollen to reconstruct past forests is that the proportion of a specific pollen type as a percentage of the total pollen in a sample may not be representative of the proportion of the particular type of tree producing that pollen in the forest. Many tree species, most notoriously pine, are greatly overrepresented in the pollen data because they produce massive amounts of pollen designed specifically for wind dissemination. Other tree species—pollinated by insects and thus with flowers and pollen not designed for easy transport by wind—are poorly represented in the pollen data. So, the general picture we get from pollen data is that the prehistoric forests of the Ozarks had a lot of oak, some hickory and pine, and enough open or thinly wooded areas to produce significant amounts of grass pollen. At the same time, we have to admit that the presence of trees such as maple, beech, and ash that do not produce abundant windborne pollen remains undetermined on the basis of pollen data alone.

Early explorers in the Ozarks and nearby areas provide a few glimpses of what the land looked like before settlement. Dunbar and Hunter explored the headwaters of the Ouachita River, in a region a bit south of the Ozarks, as an official expedition in parallel with that conducted by Lewis and Clark. They noted the abundance of shortleaf pine on uplands, with a very distinct demarcation between deciduous forests on the lower slopes and pine-dominated forests on the ridge-tops. It is likely that a similar aspect of pine-covered ridges and lower deciduous forests applied to the Ozarks, where pollen data demonstrate that pine was an important, if by no means dominant, part of the landscape. Other early reports are much less definitive than the official Dunbar and Hunter journals. There are reports by early visitors, such as Henry Rowe Schoolcraft, of relatively open woodlands with widely spaced trees where horsemen could easily pass through. One specific observation by Schoolcraft on his 1818 journey across the Ozarks is especially informative. He described some large open areas on the uplands of the Salem Plateau as similar to the lush prairies of Illinois, with two important exceptions: the vegetation was lower and sparser than that supported by the rich glacial till soils of Illinois; and the land surface, though appearing so level from a distance, was interrupted by many hidden ravines and other irregularities. Of course, the ability to put together a continuous route through grassland or open woodland does not mean that all of the forest was so open and spacious. Even so, the presence of relatively open pine forests on uplands can easily explain the abundance of grass pollen in the uppermost core from Cupola Pond.

The one really quantitative record of presettlement Ozark forests comes from early survey records. The General Land Office had sections and townships surveyed, starting in 1819, as the first step in preparing land transfer for homesteads and other uses. The surveyors identified the corners of a six-mile-square township by measuring the distance and compass heading to two or more trees located near each of the corners of the township (Figure 5.1). The diameter and species of each of the trees were identified in the records. The intent was to create a pattern that could be uniquely recognized at some future date when any survey markers at the corners had been lost. "Witness trees" were blazed with distinctive marks, and surveyors chose to mark healthy trees that were large enough to have a long expected lifetime but not so large as to have thick bark that could not be readily marked (see characteristics of old trees in Chapter 6). But these survey records also serve as a useful sampling of the forest that existed at the time. Of course, we have to depend on the ability of surveyors to identify trees by species if we want to use their records to sample the forest composition at the time of their survey. Black oaks and red oaks are notoriously difficult for even experts to distinguish. The various elm and ash species are also hard to distinguish. The best we can do is to lump these species into broad groups such as white and red oaks, elm, ash, and pine. Forest-composition analysis based on these data shows that oaks comprised about 70 percent of the presettlement forest on upland sites (Figure 5.2). Pine, hickory, and black gum were around 5 percent each, with lesser amounts of maple, ash, cherry, and chinquapin. Other tree species, such as beech, basswood, walnut, and sweetgum, were probably also around but were left out of this particular sample set because the data were restricted to the main upland oak–hickory forest where these species are rarely found today.

The statistics in Figure 5.2 say something about presettlement forest composition, but they do not provide a real vision of what those ancient forests looked like. The best we can do is to consult photographs of early virgin forests reproduced in such references as Lucy Braun's *The Deciduous Forests of Eastern North America* and Ken Smith's *Sawmill*, an account of the early logging industry in Arkansas. These photographs show forests consisting of large-diameter white oaks (Figure 5.3) that would have been ideal sawtimber. Such photographs were probably made selectively to illustrate the forest at its best, but they do serve to demonstrate what long-undisturbed forest on relatively fertile sites could be like. Old-growth shortleaf pine on

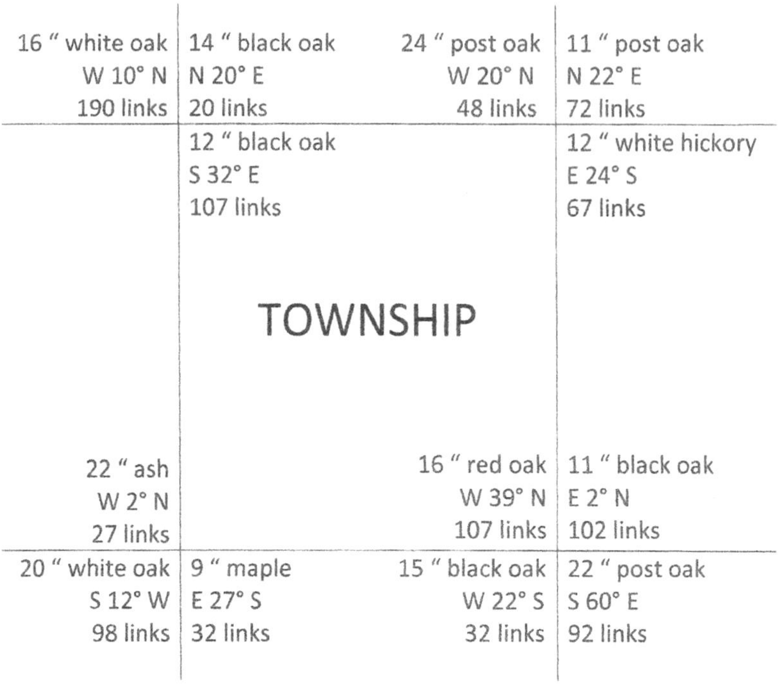

16 " white oak	14 " black oak	24 " post oak	11 " post oak
W 10° N	N 20° E	W 20° N	N 22° E
190 links	20 links	48 links	72 links
	12 " black oak		12 " white hickory
	S 32° E		E 24° S
	107 links		67 links

TOWNSHIP

22 " ash		16 " red oak	11 " black oak
W 2° N		W 39° N	E 2° N
27 links		107 links	102 links
20 " white oak	9 " maple	15 " black oak	22 " post oak
S 12° W	E 27° S	W 22° S	S 60° E
98 links	32 links	32 links	92 links

FIGURE 5.1. Schematic illustration of General Land Office records of "witness tree" locations on surveyed corners of a township in the Ozark uplands. Tree locations are given as compass azimuth and distance from the township corners to two or more identified trees of a given diameter (one link of the standard survey chain is equivalent to 0.66 feet).

upland sites occurred as somewhat open stands of large trees (Figure 5.4). The age of the trees would be indicated by the large, well-spaced branches in their crowns (other characteristics of ancient trees are discussed in Chapter 6). The open nature of these stands would have allowed for grass to grow beneath them, accounting for the proportion of grass pollen seen in the Cupola Pond record.

Early accounts also refer to barrens, grasslands, and prairies in the Ozark region, the latter mostly scattered around the periphery in the west and northwest. "Barrens" probably referred to burned-over areas on steep south-facing slopes, many of which would have qualified as glades with extensive bare, exposed bedrock (Chapter 4). Some of the prairies would have been poorly drained areas on the Springfield

Witness Tree Frequency in Percent

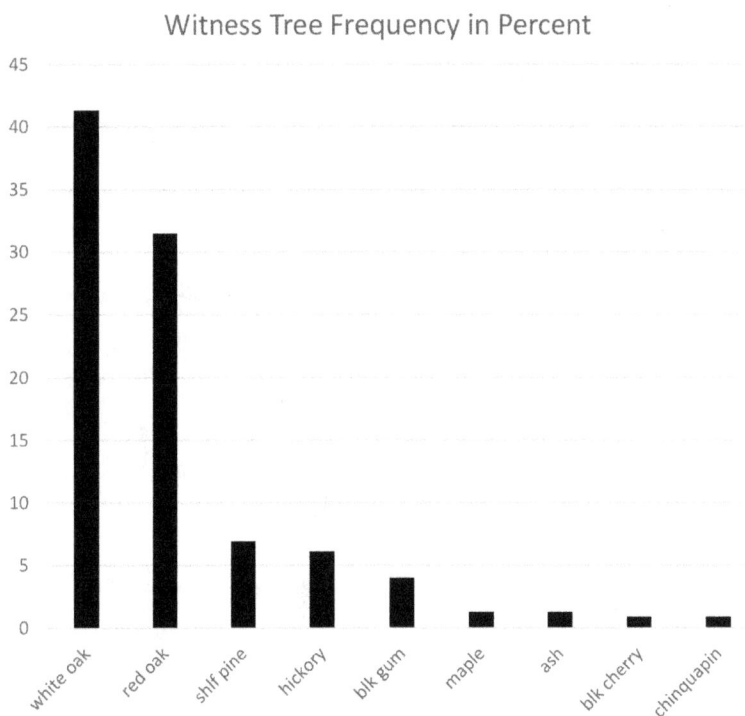

FIGURE 5.2. The composition of Ozark upland forests as inferred from "witness tree" data. The "white oak" group includes white, post, and chinquapin oaks. The "red oak" group includes black, blackjack, pin, and northern and southern red oaks (source: Thomas L. Foti, 2002, "Upland Hardwood Forests and Related Communities of the Arkansas Ozarks in the Early 19th Century," Upland Oak Ecology Symposium, USFS Report SRS-73).

Plateau underlain by clay hardpans that impeded drainage. Most of these prairie areas lie outside of the Ozark region proper. The relatively few known instances of tallgrass prairie in the Ozarks have long since been cultivated and can hardly be reconstructed. The grasslands and savannas around the edges of the Ozarks (Figure 5.5) once contained some tallgrass-prairie plant species, and the edges of these prairies sometimes had bur oak. The most common trees of Ozark savanna and prairie borders today are post and black oaks. Bur oak is found throughout the Ozark region but is not very abundant anywhere and is thus notable for its presence in this environment. Bur oak was possibly

FIGURE 5.3. Old-growth white oaks in a typical presettlement forest in the Ozarks, with the size and spacing of centuries-old oaks (as indicated by early photographs).

the only oak to remain in the area through the coldest times of the last glaciation and is the only oak species found today growing along with spruce and jack pine on the southwestern periphery of the Canadian Shield in Manitoba and Ontario, which today has a cold, continental climate similar to what must have prevailed in the Ozarks during the coldest episodes of the Pleistocene. The open landscape and frequent fire of such prairie savanna locations continued to provide habitat where bur oak could effectively compete with the many other, faster-growing oaks now present in the Ozark landscape, allowing that tree to persist as a relict species in the limited habitat available.

Consideration of the original presettlement Ozark forests eventually comes down to two closely interrelated issues: the role of fire in the landscape and the processes whereby oak–hickory forests reproduce themselves. Most oak and hickory species cannot grow well in dense

FIGURE 5.4. Old-growth shortleaf pines in a typical presettlement forest in the Ozarks, with the size and spacing of centuries-old pines (as indicated by early photographs and travelers' descriptions).

shade, and yet all evidence indicates that the prehistoric forests of the Ozarks have been dominated by oak and hickory for many thousands of years. Oak forests must have been capable of reproduction as canopy-dominant trees died of old age and disease, or they could not have remained as the major forest trees of the Ozarks. Because oaks are such an economically valuable forest resource, foresters have long been studying ways in which oak forests can be made to perpetuate themselves. The answers have proved surprisingly elusive. Various thinning and clear-cutting operations have been tried, but they have not been consistently successful in regenerating oak- and hickory-dominated

FIGURE 5.5. Savanna-like forest edges with isolated oaks (most often post and black oaks, but occasionally bur oak as shown here) around a prairie opening were common on the western periphery of the Ozarks, but otherwise such habitats were of very limited extent in the Ozark region. A bur oak's leaf and distinctive large acorn (with its fringe-like "bur") are shown at right.

forests. Some specific observations hint at the problems. Weedy growth and sprouting from understory shrubs can interfere with the success of oak seedlings. Shade-tolerant saplings established previously under the oak overstory quickly expand their foliage where undamaged by forest operations or resprout prolifically from their roots. Oaks prepare for the coming growth season by forming large terminal buds that browsing animals such as deer find especially palatable, and deer have recovered to population densities that exceed their early historical level, in the absence of large predators and human hunting over large parts of their range. All these factors simply point out that the physical conditions and ongoing disturbance regime of today's oak forests are different from those that prevailed during late prehistoric times, when oak forests were consistently able to regenerate.

With all of the facts we have today, what is our best guess as to what conditions were like in presettlement oak forests? Many ecologists think that fire, and perhaps cyclically lower populations of browsing animals, made the difference. In a relatively open forest with a clear understory, frequent fire could keep competition from shrubs and more tolerant understory trees from preventing oak reproduction. The tolerant understory competitors generally have smooth and thin bark that is easily damaged by fire. Oaks, by contrast, develop tenacious root systems that can readily resprout after the aboveground stems are killed back. Studies of fire scars on old trees indicate that fire recurred every three to five years in presettlement forests. Fire frequency may have even increased during early settlement when lands were first being cleared. Dunbar and Hunter reported that they frequently saw or smelled smoke as they journeyed up into the Ouachita Mountains in the fall and winter of 1804. Frequent fire would have kept the understory clear of competition with oaks and hickories that, with their tenacious root systems, had become established in what foresters define as "advanced regeneration" (Figure 5.6). Then windstorm or disease could provide the opportunity for these small trees to grow up to become canopy dominants, replacing the older trees. But fire suppression became the order of the day after settlement, and fire frequency was reduced to fifty years or more on most Ozark sites.

A related question is the extent to which the Native Americans had already altered North American forests, both directly through efforts to manipulate the landscape and indirectly through accidental setting of fires. Deliberate use of fire could have been part of a hunting strategy, both to make hunting easier in an open understory and also to encourage the growth of better forage for game animals. Numerous authors have argued that such landscape manipulation occurred, but specific evidence has been hard to come by. Prehistoric Native Americans have also been accused of causing the extinction of many species of large animals, such as the mammoth and the horse. The fossil record itself is so naturally sparse that it is hard to tell exactly when these animals and their large predators went extinct or if their populations had been in decline for an extended period. There is now strong evidence supporting the "overkill hypothesis," which states that human hunters arriving in the New World annihilated many large but "naive" wildlife. Large animal fossils are rare and hard to date, but fungal spores rain down continuously as layers of sediments are deposited in lakes. Among these are the spores of a specific type of fungus (*Sporormiella*) that inhabits the dung produced by large animals. The profile of these spores shows a steady level in lake sediments before abruptly falling off about fourteen thousand years ago. That pattern argues strongly for some sort of abrupt extinction, which

Woodland free of fire for many years

FIGURE 5.6. (*above and facing page*) Comparison of one forest scene, in a location protected from fire for many years, to another forest scene in an area subjected to repeated burnings (at a frequency comparable to the inferred three- to five-year recurrence interval in presettlement forests). A woodland kept free of fire has cedars seeded into the understory (**a**), thickets of blackberry and greenbrier (**b**), numerous sugar maple saplings with retained leaves in the winter (**c**), and red maple working its way into the canopy (**d**). A regularly burned woodland is composed of oak and hickory, with a relatively open understory in which a few oaks are resprouting after smaller stems were killed by fire (**e**), other oaks of basal-sprout origin are working their way into the canopy (**f**), and only the coarsest, deeply charred woody debris remains on the ground (**g**).

happens to coincide with the time when human hunters were just beginning to populate the interior of the Americas. Attempts to blame the loss of so many large animals (the "megafauna extinction") on climate change are weakened now that it is known that climate cycles as abrupt as the warming at the end of the last Ice Age have occurred many times in the past without leaving evidence of multiple extinctions. Looking at the way in which African elephants and other large herbivores affect the habitat where they live today, we can surmise that the loss of a whole range of large plant consumers would have had serious consequences for our landscape when humans first arrived. An abrupt change in the way vegetation was consumed could have

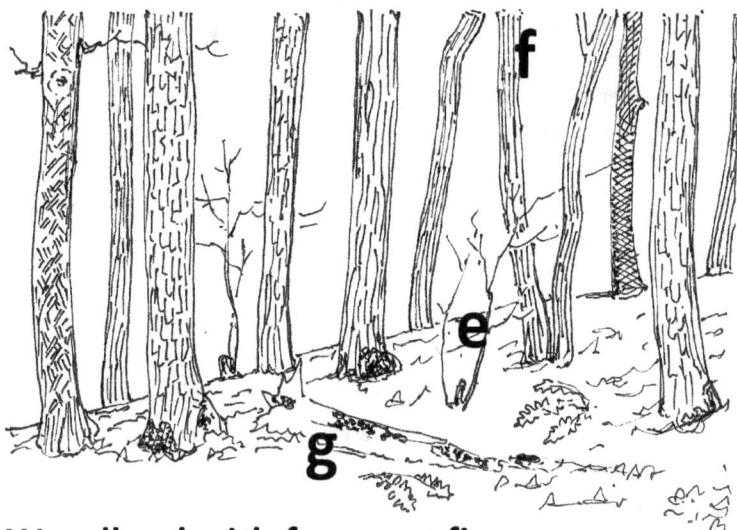

Woodland with frequent fire

produced a significant increase in the amount of burnable material (the fuel load) present on the landscape, such that human hunting had a major, if indirect and unintended, impact on the fire regime.

The increasing incidence of fire related to the presence of early humans—whether deliberate or not—may mean that to find truly virgin forests, we would have to seek out evidence from previous climate cycles with weather similar to that of the Ozarks today but without any humans involved. Pollen evidence from the American Midwest indicates that in the period from eight thousand to six thousand years ago, the climate was slightly warmer and drier than at present. Yet pine, which is considered to be adapted for dry conditions and the occurrence of at least occasional fire, was absent from the Ozarks at that time. Pine as a proportion of the forest is known to have increased throughout the American Southeast and the Ozarks after about four thousand years ago. By that time, Native Americans were practicing agriculture based on wild crops such as goosefoot and sunflower, and thus they would have already been used to manipulating the landscape at least locally. Intensive cultivation of introduced crops such as maize and beans began only after about 800 A.D. and was mostly concentrated on the fertile bottomlands of major rivers outside of the Ozark region. If the relatively recent appearance of abundant pines (see Figures 2.17B and 2.17C), even as the climate was growing slightly cooler and wetter, is a result of an increased fire frequency favoring conifer reproduction, then that increase in pine could possibly be taken as a measure of the cumulative effect of human activity on the prehistoric landscape.

Even if scientists now recognize the importance of fire in the historical ecology of our forested landscape, attempts to return fire to the forest today are confronted by several serious problems. Application of fire in "controlled burns" requires conditions that are dry enough to allow fire to spread but not so windy or extremely dry that the fire could get out of control. In the meantime, the understory population of tolerant species of trees allows even thin-barked saplings to grow to such a large size that they are difficult to kill with fire. This problem is compounded by the fact that when the fuel load has built up over a long time, the potential for a disastrous fire that threatens property adjacent to our modern forests is significant and has to be considered before applying fire to the landscape. Even then, the expansion of shade-tolerant trees into upland areas from which they were formerly excluded increases the local seed source for maples, cucumber tree, and beech, thus enabling the introduction of these trees into formerly oak-dominated forests. Red cedar, in particular, has achieved such a ubiquitous density that the species is literally out of control in our forested landscape. Low-intensity ground fires just cannot remove the presence of numerous large red cedars that have become part of the regenerating forest. There may have been catastrophic fires that destroyed great swaths of forest in the past, on the occasion of droughts more severe than any in recorded history. The Ozarks occur in a continental interior setting and just east of the transition zone between dry grassland and deciduous forests. Tree-ring studies in which tree growth rates are correlated with measured rainfall in the historical era clearly suggest that there have been such severe droughts in the past two millennia. It just might be that such occasional catastrophic disturbances are a part of the natural ecosystem of the Ozarks, and that another of these periodic "resets" may be needed to get our modern forests back into something approaching the primeval condition. The question for hikers to ponder is whether such a drastic reset of the forest environment is either necessary or desired.

CHAPTER 6

Forest Changes over Time

Hugh Raup, distinguished former director of the Harvard Forest, was once famously asked about the original forest of New England, and he famously responded, "Trees, of course." That provides an answer of sorts but effectively dodges the question, because the real issue was exactly what kind of trees. The kind of trees can indicate a lot about what has been going on in a landscape, and it was the character of the local primeval forest that was really in question. In this chapter, we look at things you can see in an Ozark forest and what those observations—of both the species present and their characteristic shapes and sizes—can say about processes that have occurred in the past.

A mature older-growth forest in the Ozarks is dominated by oaks, hickories, and a few other associated trees (Chapter 3). There will be a number of smaller trees just working their way into the canopy where individual trees have been removed by disease or storm damage. If the forest is composed of maturing second-growth trees, the number of smaller-canopy individuals may be fewer, and the largest trees may all appear to be roughly the same diameter and height. A plot of woodland more recently generated from a pasture or field will have a larger number of smaller-diameter trees, many of them varieties—like hackberry, sweetgum, walnut, elm, and cherry—that are adapted to seed into such locations. Small sections of stream bottom that were abandoned at a specific time often have nearly pure stands of trees such as sweetgum, walnut, or box elder, presumably dependent on a particular seed source that happened to be present in the area at the appropriate time. There will be tangles of greenbrier and many young grapevines draped from these younger trees. Red cedar and occasionally pine may be important components in the stand, especially in a pasture that was abandoned in stages, because these conifers are less palatable than many of the young hardwood trees to livestock occasionally let loose on the land.

The contrast between abandoned cropland and older forest is illustrated in Figure 6.1. The difference in the size, shape, and density of the trees on either side of the field boundary is evident. Note the relatively symmetrical crowns of the trees in the abandoned field and the presence of smaller dead trees in the understory that have been crowded out as the forest matures. By contrast, the established forest is dominated by oaks and contains a number of older, dead, or failing trees. Also note the small shelf or berm that lies along this edge. Repeated plowing of the field along contours of the slope causes the soil to move downhill during each plow cycle. Modern farming practices call for plowing along topographic contours. But the most fertile upland areas in the Ozarks lie on shelves underlain by shale on slopes between sandstone ledges, so it would be natural to plow along the long axis of the field in any event. Since gravity pushes overturned soil downhill, the plowing generates a little shelf at the bottom edge of the field and a corresponding embankment at the upper edge.

The Ozark region is mountainous, and cultivated land was mostly confined to rich alluvial soils along streams and small rivers or to broad shale-based benches higher up on hillsides. These rich alluvium- or clay-based soils can generate large trees in a relatively short time—often no more than fifty years. Fencing, building foundations, and the like will often show that the land was once inhabited. Stands of older trees may look old, but a truly old-growth stand will have been forested long enough for trees to have begun replacing themselves after windfall, one of the most common causes of canopy openings in such forests. This process gives what is known as a "pit and pillow" topography to the forest floor (Figure 6.2). When mature trees are pushed over by the wind, their roots tip up a mass of soil with them. After decades of this process, the decaying tree leaves small mounds (pillows) with adjacent, corresponding pits. Plowing of such soils eradicates the pits and pillows, and it would take a long time for new ones to develop after trees are allowed to reclaim the land. Long-term trampling by livestock has less of an effect but can also eventually subdue the pit and pillow features.

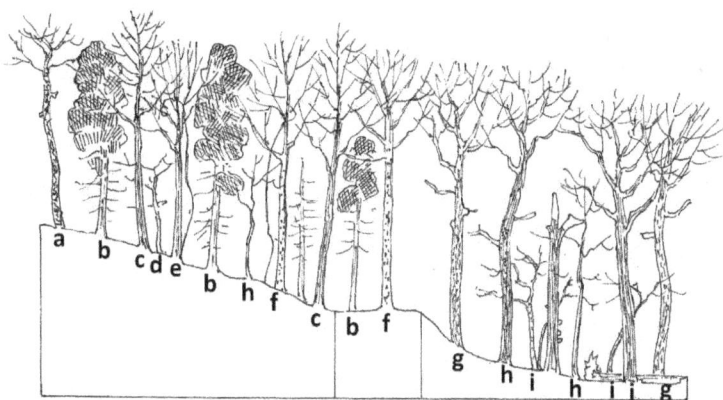

FORMER PLOWED FIELD   PLOW BERM   ESTABLISHED FOREST

TREE SPECIES KEY

a	Black cherry	f	Hackberry
b	Red cedar	g	White oak
c	Sweetgum	h	Black oak
d	Redbud	i	Dogwood
e	Winged elm	j	Northern red oak

FIGURE 6.1. Cross section of a hillside, illustrating the contact between established forest (right) and regenerating forest (left) on a former plowed field on the shallow slope of a hillside bench underlain by relatively rich soils derived from shale. The younger forest contains rapidly growing tree species that typically seed into open fields, while the established forest is dominated by oaks. Note the presence of a plow berm on the downhill side of the old field.

Sometimes a simple inspection of a plot of forest can be misleading, as was the case with the stand of oak and hickory depicted in Figure 6.3. In 2010, the overstory in this stand consisted of white, black, and post oaks, along with a few mockernut and black hickory. The largest trees were in the sixteen- to twenty-inch diameter range and distributed rather evenly among the three oak species. Of the three, black oak is generally the fastest growing, but it's also relatively short lived. Several of the largest black oaks on this plot were recently dead at the time of the study. Some white oaks were also fairly large, but we suspected that a few of the post oaks were older by virtue of that species' notoriously slow growth. Tree-ring cores were obtained

FIGURE 6.2. Old-growth forest can be identified by the presence of "pit and pillow" features (**a, b, c**): small mounds adjacent to depressions mark the places where previous generations of mature trees were blown over in windstorms, raising up large root plates that eventually decayed, creating this characteristic topography.

from all the trees shown in the figure and from many more in the surrounding forest. Two of the large post oaks turned out to be over two hundred years old, whereas all the other oaks originated in the 1930s or shortly thereafter. Thus, the interpretation is that this plot has always been in forest, but that most of the trees had been logged just before the time when many other oaks started to "recruit" into the stand. The two old post oaks did not look very different from white oaks of similar size, since they had previously grown up in a closed-canopy forest and were soon embedded once again in a closed-canopy forest.

In many cases, we can infer a tree's growth history by observing its shape (Figure 6.4). Trees growing up in a closed-canopy forest are continuously in competition with other trees, developing a narrow crown and quickly pruning lower branches. The same tree that has grown in the open will develop a broad-spreading crown with large lower branches. We often see peculiar oaks with large healed-over stubs for former lower branches, a large-diameter lower trunk, and several tall and thinner "riser" trunks rising from the top of the tree. This is a clear indication that this tree started growing in the open but was

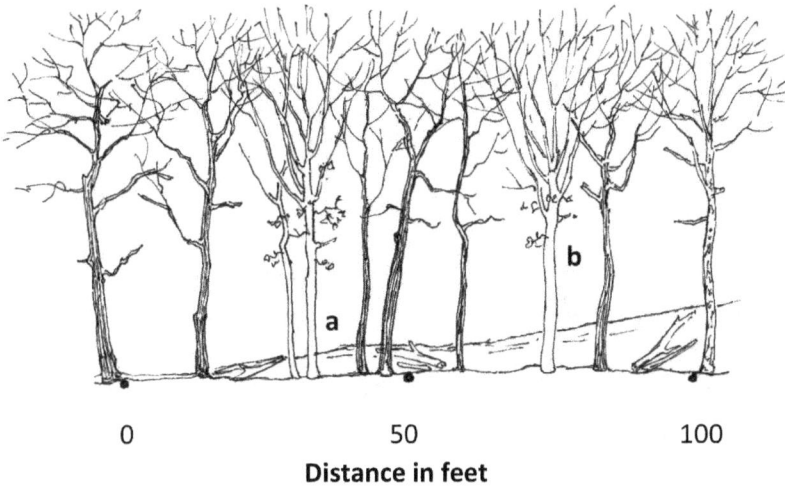

Distance in feet

FIGURE 6.3. All the trees in this forest plot appear to be about the same size, but while the two post oaks (**a, b**) are nearly two centuries old, all the other post, black, and white oaks are less than a century old.

then forced to lose its lower branches and initiate new growth in the center of its crown to keep up with competition. So the site illustrated in Figure 6.4 was once an open pasture with a large oak tree and has since reverted to closed-canopy forest. A similar scenario applies to the tree labeled "c" on the right in Figure 6.4, except that all the branch stubs are aligned on one side of the tree, indicating that it grew at the edge of a field with competing trees on the forested side of the site.

Another frequently told tree story is that of the "thong tree" (Figure 6.5). These are oddly shaped trees that bend over horizontally a few feet off the ground and then, after another short distance, abruptly turn up again. The commonly told story is that these thong trees were created by Native Americans, and perhaps by early European trappers, to point the way to a specific objective or feature. The truth is a lot less colorful. Young forest trees are often thin and supple as they try to work their way up among the larger trees around them. In the process, they will often lean outward to catch small shafts of light wherever they can. This exposes these trees to collateral damage as older and larger trees suffer wind breakage or simply die and topple

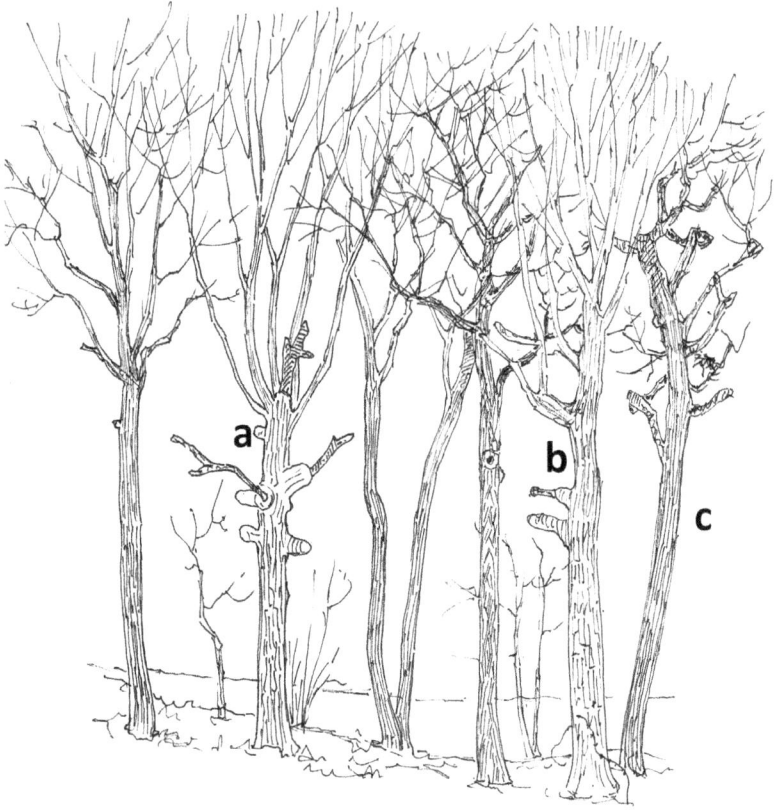

FIGURE 6.4. The shape of a forest tree can tell us a lot about its history. Here, a white oak (**a**) still bears the remains of several large, long-dead lower branches and elongated upper branches, indicating that it once grew out in an open pasture but was later forced to shed lower branches and accelerate growth in the upper crown as other trees grew up around it. Another white oak (**b**) has dead lower branches on one side, indicating that it once grew on the edge of the pasture—a conjecture supported by the presence of old barbed wire deeply embedded in the lower trunk. An adjacent black oak's (**c**) growth form is consistent with having grown up in a closed-canopy forest environment.

over as a result of disease. The saplings are prone to being bent and pinned to the ground by large fallen branches. Or it may be that an unusual episode of freezing rain causes the weight of the ice load to bend the top of this already leaning sapling down to the ground. The normal response of such damaged saplings is to start growing from a side branch that happens to be pointed near the desired upward direction. The rest of the sapling will soon die and the trunk of the young tree may spring up a bit, but the lower part of the growing tree will remain close to horizontal. With luck, the little tree will survive to become a full-grown thong tree. The basis of this bit of forest legend probably lies with the known use of "lop trees" by Cree Indians in the monotonously uniform Canadian spruce forest. These were spruce trees pruned to recognizable shapes at prominent locations, where they could be used to identify intersections on canoe routes, pointing to portages or designated tributaries along the way.

We have already discussed the ways in which oaks and other trees struggle to find their way into the overstory of a forest. Often this is a process of establishing saplings in the understory that manage to persist until an opportunity arises for rapid growth into a newly created opening. Studies of mid-tolerant species such as northern red oak have shown that it may take several such partial release events to finally give the tree a dominant position in the forest. Forest ecologists now believe that release events are essential in the natural regeneration of oak-dominated forests. Inspection of cross sections of logs along forest trails can sometimes tell the story of how large trees eventually gained their position in the surrounding woodland. Unfortunately, most Ozark forests do not have a very interesting tale to tell. Much of the area was heavily logged about a century ago, and the regenerating forest shows oaks and hickories with a simple record of initial rapid growth for maybe three decades and then slowly declining increment widths into old age. However, a few tree species have reproductive strategies that are finely tuned to the suppression-and-release cycle. One of the trees in our area that behaves in this way is the Ozark chinquapin (Figure 6.6; not to be confused with the chinquapin oak), which apparently has the ability to survive as a heavily suppressed sapling under adverse conditions while being programmed to rapidly grow upward if light conditions temporarily improve. The cross section in Figure 6.6 shows a small chinquapin tree cut for trail clearance: an inner cylinder of excruciatingly slow growth was followed by rapid growth, producing a small tree in less than ten years. Unfortunately, this tree also began to encroach on a forest trail, which led to its cutting by a trail maintenance crew. Dramatic demonstration of the ability of

FIGURE 6.5. The making of a "thong tree": an initially suppressed and slightly leaning white oak sapling (**a**) is bent and then partially broken by ice loading as a result of freezing rain and begins rapid upward growth from a small lower branch and by sprouting from a point just below the injury, in the opening created by storm damage (**b**). A few decades later, the oak has become a typical thong tree developed entirely through natural processes (**c**).

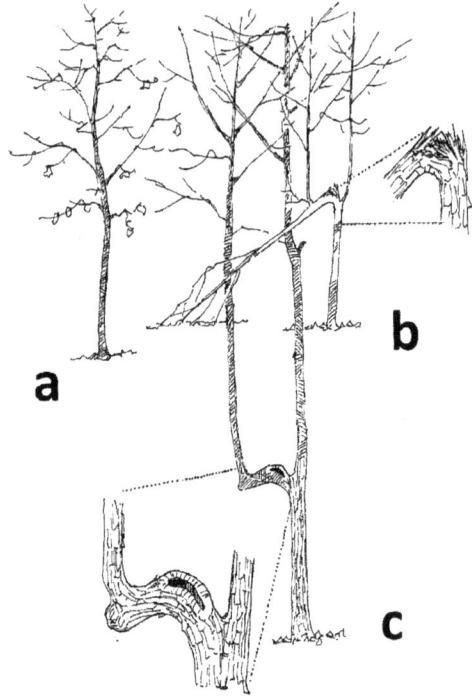

the chinquapin to respond to release is provided by examples of direct observation of vigorous growth in response to recent storm events, in which new shoots from broken branches or from the base of the tree could grow as much as six feet in a single year.

Although many of our oak and hickory species rely on release of what foresters call "advanced reproduction" to gain a position in the canopy, there are other species that can prosper in the understory of an established forest without the help of any external events at all (Figure 6.7). In the absence of fire, these trees can seed out from their protected habitats in moist ravines to colonize the upland forest. In this process, known as "mesification," upland oak–hickory stands are being invaded by moisture-loving (mesic) tree species. This process is most apparent in the height of winter, because beech and sugar maple saplings retain a good portion of their brown and shriveled leaves until they are forced off the branches by swelling buds in the spring. All three of the mesic species shown in Figure 6.7 have relatively thin

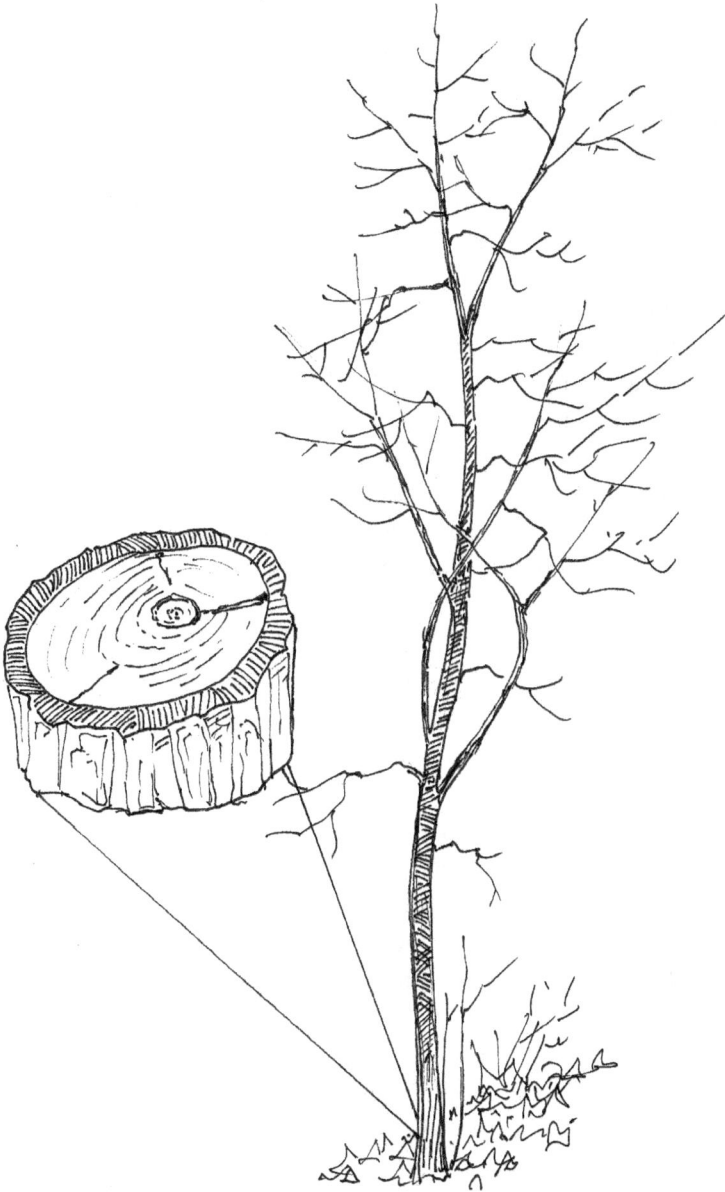

FIGURE 6.6. Cross section taken from an Ozark chinquapin cleared during trail maintenance shows an inner core of heavily suppressed growth where a diameter of about one inch was achieved over thirty years, followed by a release during which the tree grew to six inches in diameter in just eleven additional years.

bark that is easily damaged by fire and are not especially good at generating new sprouts from any root stock that survives below the bed of the fire. For this reason, many ecologists believe that the mesification we see today is a result of a long history of fire suppression in our woodlands.

Another bit of information repeatedly provided by tree-ring studies is that the size of a tree is often unrelated to its age. Many ecologists had long mourned the loss of pristine virgin forest in the most heavily populated portions of America. The limited number of such remaining stands was associated with a few carefully preserved locations such as the Joyce Kilmer Grove in North Carolina and the Hearts Content area in Pennsylvania. But tree-ring studies have shown that there are actually virgin old-growth blocks of forest even in developed suburban areas. These are bits of forest established on rocky outcrops, in swamps, or at other locations where the timber was never of any economic value and on sites that were otherwise unsuited for development. Such pieces of virgin forest have been effectively hiding in plain sight. The casual hiker would not have the ability to verify the existence of such really old forest stands by extracting tree cores, but there are other ways of identifying trees that may be especially old. Two specific characteristics can be used in this regard. One is the appearance of the bark. As trees age, their rate of growth systematically declines. This means that the bark cambium layers do not have to regenerate as often to expand the outer circumference of the tree. Bark plates become larger, thicker, and more exaggerated (Figure 6.8). Hikers can be on the lookout for trees that have such unusual bark. Not only are they interesting to see, but they may represent a long span of forest history.

The second way to identify ancient trees hiding in plain sight is to look for evidence of multiple rounds of crown damage that would have accumulated over centuries of exposure to wind and ice storms. Figure 6.9 shows a tree with a broken crown and missing branches, indicating a long career of wind battering and forest survival. That appearance is confirmed by the corresponding appearance of the bark. In general, northern red oak has extra-large and thick bark ridges (Figure 6.8), and post oak has a pronounced spiral pattern to the smaller bark plates on its trunk. Tree cores were never extracted

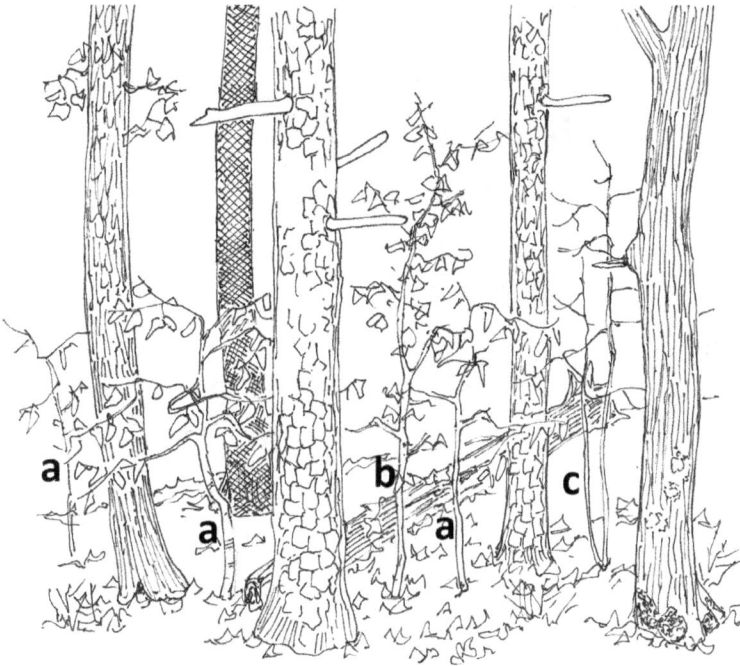

FIGURE 6.7. Mesification in action: a ridgetop forest of oak and shortleaf pine has an understory filled with saplings of mesic tree species, mostly beech (**a**) but also some sugar maple (**b**) and cucumber tree (**c**). This is a midwinter scene in which beech and sugar maple become conspicuous because they retain some of their shriveled and dried leaves in place until spring arrives.

from either of these trees, but the combination of crown damage and bark appearance provides assurance that they are probably several centuries old.

Further demonstrating that size alone is not a good indication of a tree's age, some of the fast-growing, early-succession tree species such as sweetgum and American elm can reach impressive diameters within half a century. And, conversely, a tree-ring study conducted on recently acquired parkland in Fayetteville, Arkansas, showed that trees in a stand of rather ordinary-sized chinquapin oaks were as much as three centuries old (Figure 6.10). The steep, rocky, and otherwise inhospitable site where the trees were growing, the obviously unsuitable shape of their twisted trunks for lumber, and the pattern of past crown damage all led to the study, which proved that there is,

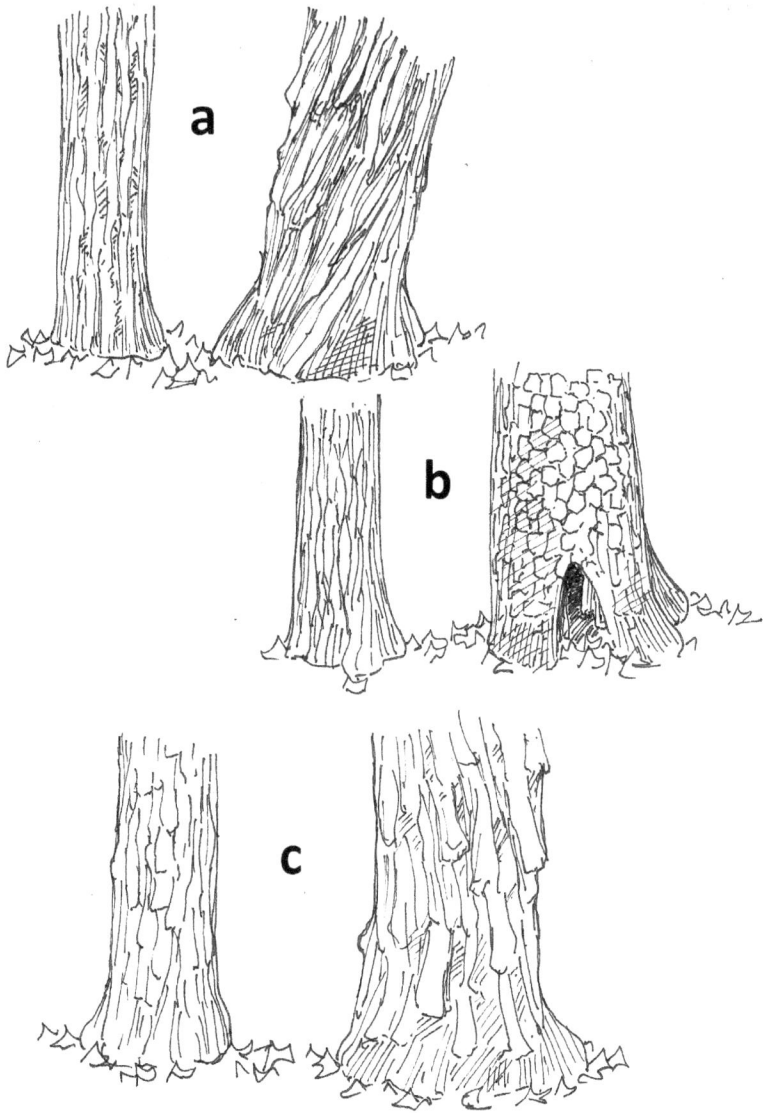

FIGURE 6.8. Bark appearance on mature (left) versus very old (right) individuals of (a) northern red oak, (b) black gum, and (c) sugar maple.

FIGURE 6.9. This ancient post oak shows the accumulated episodes of crown damage and spiral pattern of bark that are useful indicators of great age in trees.

indeed, old-growth forest within the Fayetteville city limits. And, in some cases, simple use of tree examination to estimate age can yield unexpected results (an example is the tiny Ozark chinquapin tree in Figure 6.11).

FIGURE 6.10. Ancient trees often hide in plain sight, like this modest-sized chinquapin oak growing on the upper slopes of a rocky ridge inside the city limits of Fayetteville, Arkansas. Poorly formed trees like this were overlooked by loggers, but their great age is suggested by the accumulated crown damage and the long-healed seam in the base of the trunk. Cores taken from the area show that many of the chinquapin oaks on this ridge are more than two centuries old.

FIGURE 6.11. This example of a heavily suppressed Ozark chinquapin sapling indicates that size is not always a good indicator of age, because this little tree was less than three inches in diameter but had the characteristic thickly ridged bark of a mature tree—and so was perhaps more than fifty years old—when it was identified in 2012 at Roaring River State Park in Missouri.

A good way to summarize this discussion of the changes that occur in the forest and the ways to track them is to consider the standard model for forest stand regeneration in an Ozark deciduous forest (Figure 6.12). This is a framework that professional foresters use to deal with forest management plans, but it also provides a useful way of understanding the forest structure that you see while walking through Ozark woodlands. The first stage in forest establishment is

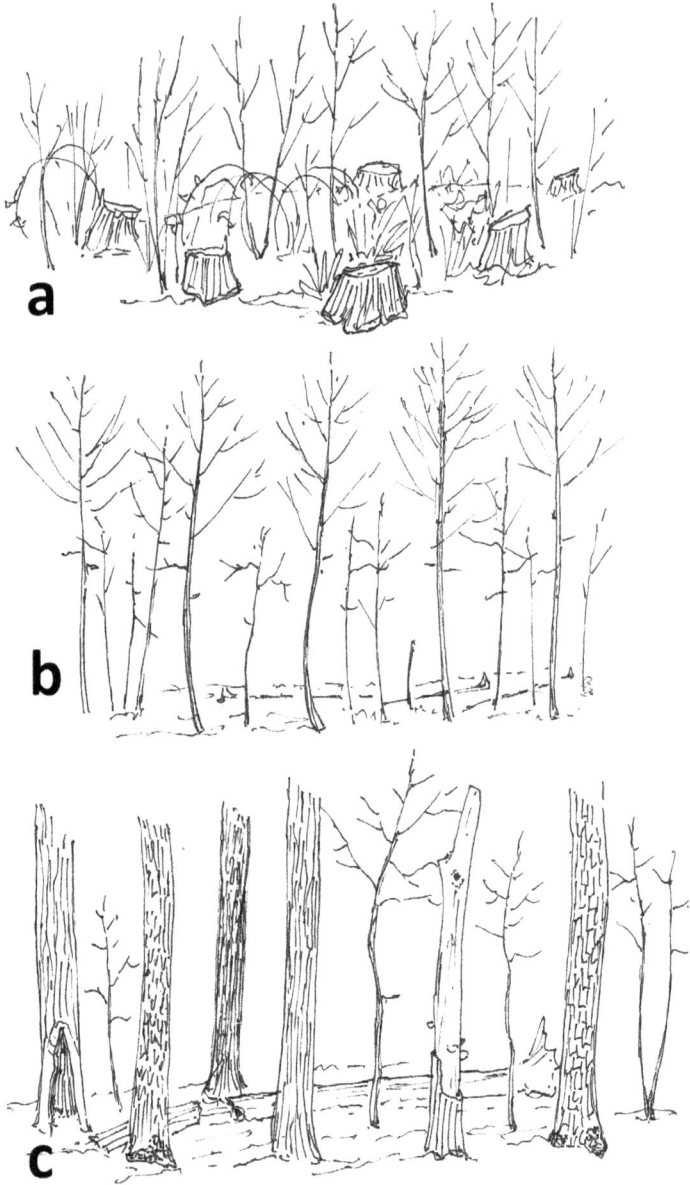

FIGURE 6.12. Forest regeneration after logging or other major disturbances is generally considered to occur in three phases: (**a**) the initiation stage, in which numerous new stems are established by "seeding in" from fruits of trees in adjacent forests or by sprouting from root stocks already present; followed by (**b**) the exclusion stage, in which the number of stems per unit area decreases as the fastest-growing trees eliminate their competitors; and finally (**c**) the complex stage, in which openings start to appear in the established canopy that allow younger trees to grow within the stand.

the initiation stage, when a major disturbance has opened up the land-scape. This is a time when trees either seed into a newly opened site or develop by sprouting from the roots of smaller, established trees. In forestry management plans this would be a logging operation, but it could also represent a windstorm or severe fire. This scenario is widely applicable to much of the Ozark region because almost every part of the region has forests that regenerated after logging about a century ago. After about twenty years, the exclusion stage begins, when no new trees are being established and the existing trees begin to exclude competitors. After about seventy years, the stand enters the understory reinitiation stage, when light gaps start to allow new tree seedlings to become established in the understory. After more than a century, the stand enters the complex stage in which old, mature trees are joined by younger trees that have grown into gaps produced as trees were eliminated by disease and wind. An important factor here is that the remaining trees have become so large that the loss of an indi-vidual tree to "windthrow" (uprooting and overthrowing by the wind) opens up a canopy gap that cannot be filled simply by the expansion of adjacent trees. Once this state is achieved, the forest might remain in this condition indefinitely. Tree-ring studies of events in some of our remaining virgin hardwood forests have confirmed that the steady attrition of trees by windstorm loss and gap replacement can actually account for stand history as reconstructed from detailed tree-ring studies. Most of the oak–hickory forest we see in the Ozarks today is so far advanced in regeneration from logging in the first decades of the 1900s that it has reached the complex mature stage. But some former homesteads and abandoned pastures, such as those illustrated on the uphill side of Figure 6.1, are still in the exclusion stage. The initiation stage can occasionally be seen in small openings produced for wildlife enhancement or in recently abandoned farmland around the edges of small urban areas located at the fringes of the Ozarks.

CHAPTER 7

Catastrophic Events in the Forest

In the previous chapter, we dealt with observations that reveal the "slow-motion" changes occurring over time in a developing forest stand. Other significant changes can occur relatively quickly—even in an instant, as when a sudden event causes catastrophic damage to individual trees or groups of trees. When you find the corpse of a tree on an Ozark outing, see if you can recognize what has happened from the evidence on hand. First, it's often important to ask whether a tree simply died and fell over under its own weight or was brought down by some other agent (Figure 7.1). A decade or more after a treefall event, the distinction may not be obvious. "Deadfalls" are trees that have died in a standing position and then fallen over, under their own weight, after their roots have mostly rotted away. They topple in various directions except on steep slopes, where most trees will have been leaning downhill already and will fall in that direction. The main indicator of a deadfall is that the tree did not carry a "root plate" of fine roots and soil with it when it came down (Figure 7.1, a, b). Trees blown over by a windstorm will often all lie pointing in the same direction, usually to the southeast or east in the Ozarks, because our storms typically come from the northwest. A blowdown most often occurs when the tree's crown is full of leaves, because that leafed-out crown offers much more resistance to wind than a bare crown. Thus, trees blown over in storms often display dried brown leaves attached to their branches for several years after the event (Figure 7.1, c). Another sign of a windthrow is the raised-up root plate, which eventually leaves a mound of soil at the base of the deteriorating log, showing that it was blown over by a strong storm and not killed by some pathogen (Figure 7.1, d). As discussed in Chapter 6, windthrow is one of the most important processes shaping the old-growth forest, and development of a "pit and pillow" topography on the forest floor is one useful indicator that this process has been going on for a long time.

Another common catastrophic event (at least for a tree) is logging. Recent logging is always shown by the presence of flat-topped stumps, indicating that trees were cut with a saw. Even decades later, when stumps have been badly deteriorated, you can see indications of logging roads and of slash piles that resulted from the removal of branches and tops from sawlogs. However, there is one distinctive indicator that often removes all doubt. When equipment known as "skidders" drag logs along temporary lanes constructed to transport timber, damage to trees on either side of the temporary road can occur, including basal scars on the trunks of trees (Figure 7.2). Fire damage can produce similar scars, but generally on a consistent side of trees and not lined up on separate sides of old roads. Such paired scars on trees are a clear indication of past disturbance by logging. Sites subject to logging are also useful in that the flat saw cuts allow you to determine what trees were present, by providing a visible cross section of the wood even after the bark has fallen away. In general, hardwood stumps rot from the inside out, whereas pines rot from the outside in (Figure 7.3). Deciduous trees are divided into two classes: those that have distinct rings of large pores (ring-porous trees) and those that do not (diffuse-porous trees). The former include oaks, hickories, elms, beech, and ash; the latter class is represented mostly by maples in our area. Furthermore, oaks are characterized by distinctive rays that cut perpendicularly across the tree rings on an oak stump. In the case of badly decayed oak stumps, where it is just possible to recognize the flat top of the stump, the rays often stand out like thin, hard wafers from the fibrous mass of decayed wood around them. Another useful diagnostic feature in old stumps and fallen trees is that the knots at branch nodes on pines are especially resistant to decay. Knots at the locations where lower branches were shed early in the tree's life tend to persist in the core of the stump as it decays from the outside in. A large pine log in the late stages of decomposition is a dramatic sight. Much of the main trunk will have decayed away, leaving an evenly spaced set of rounded knots with laterally projecting branch stubs that resist decay by being kept off the ground (Figure 7.4). The pine "skeleton" that remains bears an uncanny resemblance to the backbone of a dinosaur.

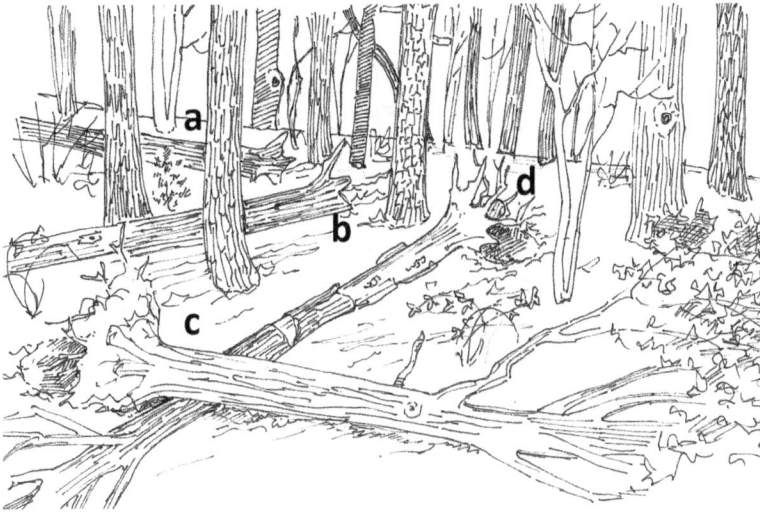

FIGURE 7.1. Fallen trees on the forest floor can be either "deadfalls" (**a, b**), the term for trees that died while standing and whose decaying trunks fell over when the roots rotted out; or living trees blown down by a windstorm (**c, d**) that tore out root plates with soil and rocks, leaving a substantial pit at the base of the tree. Note that tree **c** has a crown full of dead leaves, indicating that it was blown down during a summer storm. Tree **d** was blown down earlier: it lies in a way that indicates a different wind direction and shows somewhat more advanced decay. Note the substantial stone still embedded in the deteriorating root plate of the latter tree.

One of the most dramatic events in the forest is a lightning strike. In theory, lightning will strike the topmost point on a ridge because that is the shortest path between cloud and ground. But it rarely works out that way. There is a good deal of randomness in cloud convection, and lightning strikes don't follow a simple, direct path. For one thing, the current path is created by literally tearing air molecules apart, and the tear starts where the electric field gains "leverage" near a sharply pointed object (Ben Franklin discovered this with one of his inventions, the lightning rod). Unfortunately for trees, they usually have pointed leaf tips that serve as lightning attractors. Trees are damaged in lightning strikes by the massive conduction of electrical current through their vascular system. The tree's sap boils, and the resultant steam explosion can turn the trunk into a pile of splinters—clearly an event not survivable by the tree (Figure 7.5, a). More often, the bark and other outer layers of a tree struck by lightning will simply split

FIGURE 7.2. Forest scene showing paired scars at the base of two trees, caused by movement of logging equipment in the narrow space between the trees; water-filled hollows also indicate old wheel ruts from forestry equipment.

violently open and leave a spiraling gash in the side of the trunk. The tree can often survive this, but the scar can be apparent for the rest of its life (Figure 7.5, b).

Ice storms are another common source of tree damage that can happen in a single day and have effects that persist for many years. Our area is prone to such damage because it is poised in a continental region where sharply different air masses regularly override one another. Normally, water drops from tree branches before it has time to freeze, even when the air near the ground is below freezing. However, in the fortunately rare situation called "supercooling," the air mass on the ground becomes extremely cold, and droplets falling from warmer rain clouds above are cooled to a temperature well below freezing. But the droplets don't freeze right away; to form ice crystals, they need some kind of "push"—from an impact with a tree branch, for example—and then ice forms instantaneously. In a supercooling event, inches of ice can coat every branch and twig over a large area. As the ice load accumulates over time, a very distinctive form of damage

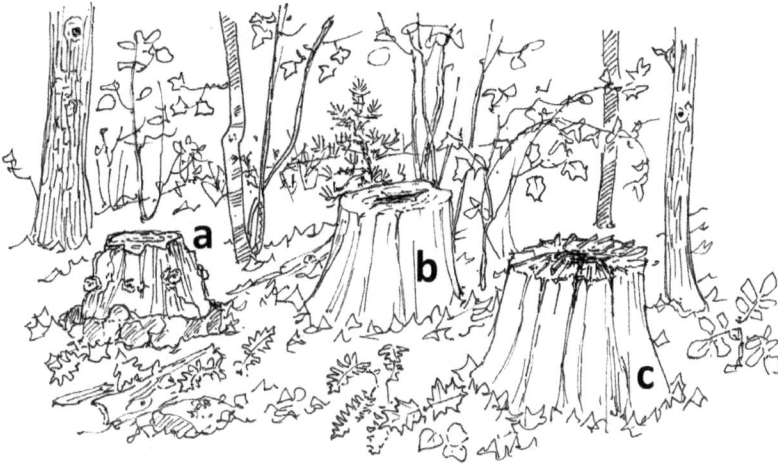

FIGURE 7.3. The flat tops of these weathering stumps indicate that the trees were cut in a logging operation. Conifer (pine) stumps (**a**) deteriorate from the outside inward, exposing knots from former low branches, with fragments of pine bark visible in litter in the forest floor at the base of the stump. Hardwood stumps like American elm (**b**) deteriorate from the inside out; and deteriorating oak stumps (**c**) display erect "fins" at the top that represent hardened wood rays.

FIGURE 7.4. Decay of pine logs leaves distinct lines of woody knots from branch nodes on the trunk, resembling a dinosaur's backbone.

FIGURE 7.5. Two examples of lightning damage to forest trees: (**a**) a black cherry, completely splintered by a steam explosion created by a lightning strike; and (**b**) a spiral scar left on a large post oak by a lightning strike that occurred decades earlier.

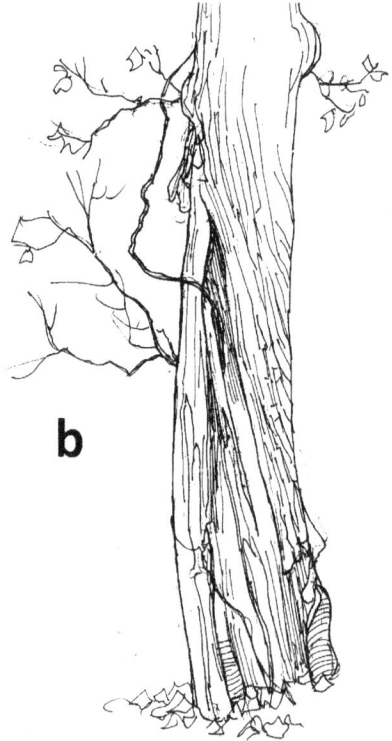

can occur as trees gradually bend over in a slow-motion kind of failure. Instead of breaking off or tipping the entire tree over, the trunk progressively splinters (Figure 7.6). Abrupt breakage and tipping of root plates also occur during ice storms, and many broken branches will be left hanging in tree crowns as potential "widow makers" for years afterward. But a predominance of bending and shattering in a damaged forest is a good indicator that the causal agent was ice loading and not storm winds.

Tornados are yet another source of damage to which the Ozark region is unfortunately prone. Most wind gusts are subject to turbulent mixing, which causes a coherent gust to soon disintegrate into a disorganized series of wind eddies. Tornados (along with upper-atmosphere jet streams) are a notable exception to this rule because rotation can effectively suppress the turbulent process, allowing local winds in excess of two hundred miles an hour to rage over a tightly confined area. Damage to the surroundings outside the exact width of the funnel can be surprisingly minimal, because rotational stabilization keeps the tornado's energy tightly focused. Tornado damage

FIGURE 7.6. Splintered oak showing the signature of ice-storm damage: the steady buildup of ice caused the tree to lean, bend, and then splinter as the weight of the ice load slowly increased and tore the trunk apart.

to forest trees is not subtle—there will be an alley where every single tree is simply snapped off five or ten feet above ground level, with virtually untouched trees nearby (Figure 7.7). Many of the lower tree trunks that remain standing will have their bark scoured away by the abrasive cloud of fine debris driven by tornado winds. There may even be pieces of foreign debris, such as aluminum gutter sections or trailer siding, lying on the ground or suspended in adjacent treetops at locations far from any human habitation.

Fire has always been a part of Ozark forests, although its frequency has apparently changed over time (see Chapter 5). There are two basic methods that scientists use to investigate the past occurrence and recurrence intervals of fires in a forested location. These are (1) the presence and distribution of microscopic charcoal particles in the soil and (2) the presence of basal fire scars on trees. The analysis of soil charcoal is a complex process and is best left to the specialist,

FIGURE 7.7. Tornado damage can completely remove the crowns and upper trunks of trees when the tornado is severe, causing much more damage than the typical windstorm. Note the trunks with bark removed by abrasion from flying debris, and the presence of undamaged trees in the near background—the tightly focused vortex of the tornado cuts a relatively narrow and well-defined path through the forest.

because interpretation is difficult and fraught with many uncertainties about transport and depositional processes. Fire scars are much more straightforward, and you can readily identify them along forest trails. They are triangular areas at the base of trees where the heat of a fire has killed the bark and cambium. These scars are almost always located on the uphill side of the trunk, where debris tends to build up over the years. The fire burns hottest at these locations, accounting for the dead tissue on that side of the tree (Figure 7.8, a, b). Cross sections of trees with fire scars can be used to date the occurrence of fires by counting the rings back and noting the number of individual scars that extend between layers of callus tissue growing over the edge of the wound after each event (Figure 7.8, c). Of course, not all fires will scar all trees during any single event, so there is a bit of statistical interpretation involved in the use of fire scars to measure fire recurrence intervals in the past. But to get a feel for how often fire has occurred in a particular area, a hiker can simply note the presence of fire scars on trees as an indication of past burning and can look for any available stumps along the trail on which multiple fire scars can be observed.

A relatively minor disturbance that can have an important effect on a few otherwise well-formed tree saplings is the "buck rub" (Figure 7.9). This term refers to extensive damage to the base of a sapling, created when young whitetail bucks rub the velvet covering off their maturing antlers as the fall season begins. This can permanently deform what otherwise might have become a straight and vigorous young tree. The total effect of buck rubbing on the forest is probably negligible, but the rubs themselves can be an interesting puzzle for the casual hiker who is not familiar with the phenomenon. Bucks seem to find that the smooth bark texture and spring-like resistance of a young red maple, black cherry, or Ozark chinquapin tree makes for an especially satisfying rub experience.

Another minor but interesting natural artifact you can see while hiking in Ozark woodlands is the pattern of holes drilled in the bark of trees by sapsuckers (Figure 7.10). These woodpecker-like birds drill holes in many varieties of trees, including red oak and serviceberry. They drink the sap that collects in the holes, and in the process they sometimes also consume living cambium tissue. The holes occur in striking geometric arrays; except for this unexpected symmetry, the

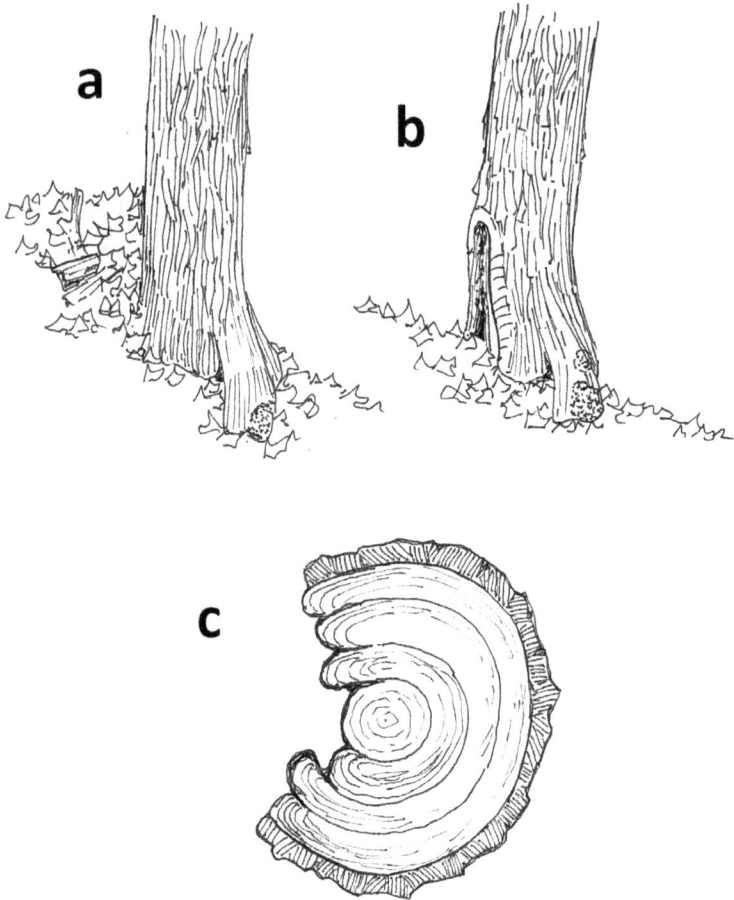

FIGURE 7.8. When debris that has collected on the uphill side of a tree's base (**a**) burns, the extra-hot fire damages the bark on that side, leaving a fire scar (**b**). A section from the trunk of a tree that has experienced a series of such fires (**c**) can be used to determine fire frequency by counting the number of annual growth rings in the wood that grows over the edge of each individual charred surface along the edge of the scar.

bark looks like it's been strafed with a machine gun. In most cases, sapsucker damage has no major effect on forest trees, but sapsuckers can sometimes seriously damage or even kill isolated landscape specimens. An even more interesting sight is what happens to the sapsucker-drilled tree in the years after the holes are made. The tree continues to grow and cambium heals underneath the holes. The expansion of

FIGURE 7.9. When the peeling velvet on a male whitetail deer's maturing antlers causes the buck to vigorously rub those antlers on a young tree that gives just the right amount of spring-like resistance, a relatively minor injury can have important implications for otherwise well-formed young saplings that would seem to have a prosperous future. Saplings of red maple, black cherry, and Ozark chinquapin (shown here) between one and two inches in diameter seem to provide just the right bark texture and resistance to result in a "buck rub."

the tree trunk then causes the holes to expand in such a way that an intricate latticework of outer bark is created that looks almost like lace—a truly unique sight to observe in the forest.

Tree-ring analysis is one of the more sophisticated ways to obtain a detailed history of a specific forested study site. That technology is beyond the scope of this book, but a basic appreciation of the science involved is useful for any forest observer—just as the average TV homicide detective has no real knowledge of organic biochemistry but still understands the value of DNA analysis at a crime scene. The discussion here is just a way of providing basic tree-ring-analysis literacy. In tree coring, pencil-sized cores are extracted from trees, mounted on wooden blocks, and polished with sandpaper to make the rings visible. Then ring widths can be sequentially measured using a specially designed wood microscope that puts crosshairs on each ring boundary to precisely measure growth-ring thickness. The thickness of each ring in the series represents the growth capacity of that tree in that specific year. Variations in ring thickness can then be used to identify events that affected the life of individual trees. The ring-width series from many trees at a location can be summed to average out factors such as competition and windstorm damage to individual trees, thereby providing a series that represents the effects of climate (primarily rainfall) on the forest. Scientists have used sophisticated analyses of tree-ring data to reconstruct the record (known as a "tree-ring chronology") of drought and flood events using these average ring series that extend far into the past, beyond the relatively short interval of recorded meteorological data.

Quantitative tree-ring analysis is best left to the specialist, but we can frequently see cross sections where trees have been cut by logging or where fallen trees have been cleared from trails (Figure 7.11). Counting the rings can provide an indication of the age of trees in the local forest. You can also look for abrupt changes in growth rate that represent past events in this specific location that allowed this individual tree to experience accelerated growth by elimination of competition, or reduced growth caused by damage to its crown. Careful attention to such information can provide some insight into the history of the forest in relation to "release events" and storm damage. In specific cases, interesting changes in growth rate can be com-

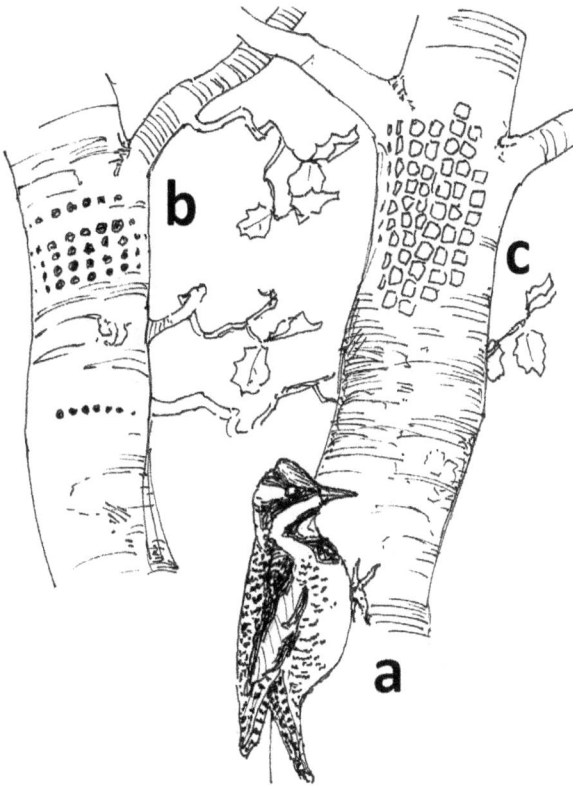

FIGURE 7.10. Sapsuckers (**a**) drill arrays of evenly spaced holes (**b**) in the bark of trees, and the expansion of the bark in future years causes these drinking holes to develop into an intricate and interesting lace-like pattern (**c**).

pared to averaged tree-ring series, which are generally available in the literature, to determine whether these changes are attributable to climate extremes such as recovery from drought or to a local event that affected this single tree. In Chapter 13, we discuss chestnut blight and the loss of mature Ozark chinquapin trees from our forests, a case in which this technique was used to determine the time when blight first arrived in the Ozarks (also see below).

When logs are beginning to decay and bark is no longer present to indicate the species, wood anatomy can sometimes indicate the species or at least narrow the possibilities. Many hardwood trees have distinct rings of large vessels generated at the start of the growing season (ring-porous species, as discussed above; Figure 7.12, a, b). Since these vessels are generated using stored resources from the previous growing season, they are often very regular in appearance. By contrast, the much

FIGURE 7.11. Although sophisticated tree-ring analysis requires a special wood microscope and associated digitizing and analysis equipment, the untrained Ozark hiker can learn a lot about a forest site's history by examining the exposed cross sections of logs that have been cleared along walkways (as here, along the Ozark Highlands Trail).

narrower vessels produced during the growing season determine the thickness of the complete growth ring, with cambium growth shutting down earlier in years of unfavorable climate. Oaks (Figure 7.12, c) are ring-porous trees but also have radially extending layers of wood known as "rays" that cut across growth rings. Diffuse-porous trees such as conifers and some hardwoods (Figure 7.12, d, e) have rings consisting of narrow vessels, such that the rings are simply separated by a kind of seam. Some of these diffuse-porous trees have larger vessels embedded among the narrow ones (Figure 7.12, d), but these are randomly dispersed and are not organized into distinct rings. Specialists with a wood microscope can identify individual tree species by examination of their wood, but the five distinctions shown in Figure 7.12, a knowledge of what trees are likely to have grown in a certain habitat, and some badly preserved bark fragments on or near the log or stump are often enough to let hikers make a good estimate of the identity of long-dead trees they may encounter on their forest outings.

One important aspect of tree-ring appearance is an asymmetry in which the growth rings are dramatically wider on one side of the trunk (Figure 7.13). This occurs when trees are bent over and the trunk attempts to grow itself back into an upright position. Logic

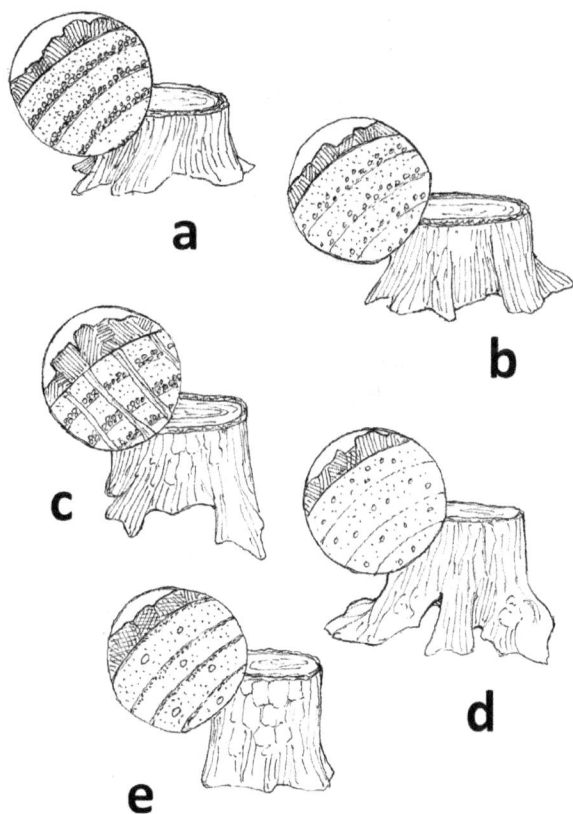

FIGURE 7.12. Tree-ring structure and characteristic appearance of wood in (**a**) white ash, a ring-porous hardwood; (**b**) black walnut, a semi-ring-porous hardwood; (**c**) northern red oak, a ring-porous hardwood with rays; (**d**) sugar maple, a diffuse-porous hardwood with randomly distributed large-diameter vessels; and (**e**) shortleaf pine, a diffuse-porous conifer with a few large resin ducts and a relatively dark band of "latewood" formed at the end of each growing season.

would suggest that all trees use the same method to accomplish this, but that is not the case. Hardwoods such as oaks and maples try to pull themselves upright with thickened growth on the uphill side of the bend. By contrast, conifers like shortleaf pine and juniper push themselves upright with thickened growth on the downhill side of the bend. In both cases, the transition from symmetrical to nonsymmetrical rings occurs outside the annual ring produced at the time when the injury occurred. This can be used to date the time when the trunk was bent by an ice storm or landslide event (Figure 7.13, c). In places like California, where there are long-lived trees (redwoods) and major landscape disruptions (earthquakes), this kind of tree-ring dating has important applications. Although the method is used by forest ecologists in Arkansas and Missouri for specialized studies, it is fortunately less important in the Ozarks, where our major seismic hazard (the New Madrid seismic zone) has been shown to have an approximately five-hundred-year recurrence interval for major events, which exceeds the life span of almost all our native tree species. Given that the last major New Madrid earthquake occurred in 1812, we can expect to be spared another such event for a few more generations.

Sometimes inspection of the ring structure of log sections encountered along an Ozark forest trail can tell a more complicated story than just regeneration after the heyday of regional timber production in the early part of the twentieth century. Figure 6.1 (in the previous chapter) shows the contrast between intolerant trees developed on a long-abandoned field and an adjacent, established oak forest. Note that there is a relatively small black oak embedded in the old-field forest. Red oaks often bide their time in situations where they cannot compete with faster-growing trees such as sweetgum and hackberry, but they begin to expand upward as those larger trees enter old age or encounter storm damage. According to forest ecologists, this is how oak–hickory forests eventually turn over in the normal process of forest succession (Figure 7.14). Intensive forest forensic studies at the Harvard Forest in Massachusetts have certainly demonstrated this process in the case of northern red oak, a New England tree that is a major component of Ozark oak–hickory forests. Since there are usually many sectioned logs cleared from the typical Ozark forest trail, it is interesting to see if you can recognize how many release events

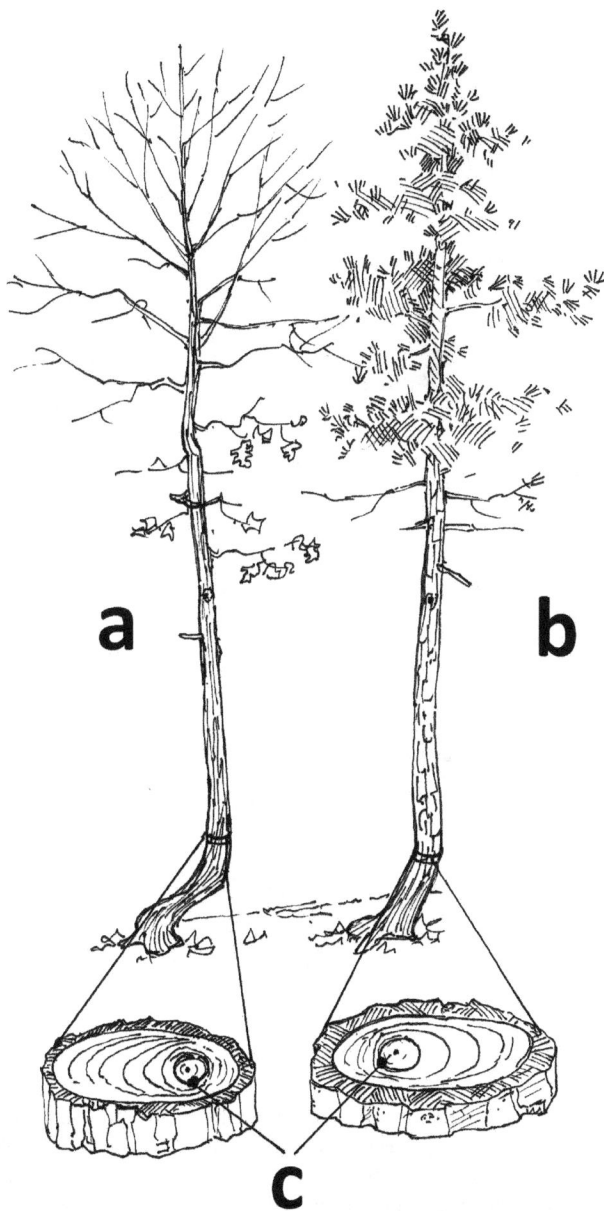

FIGURE 7.13. Schematic illustration of reaction wood formation as a tree resumes growth after being bent over by ice-storm or wind damage. Deciduous trees like white oak (**a**) pull themselves upright with thickened growth rings on the upward side of the trunk, whereas conifers like shortleaf pine (**b**) push themselves upright with thicker growth rings on the downward side. In either case, the date on which the damage occurred can be determined by examining growth rings (**c**) in a core or cross section.

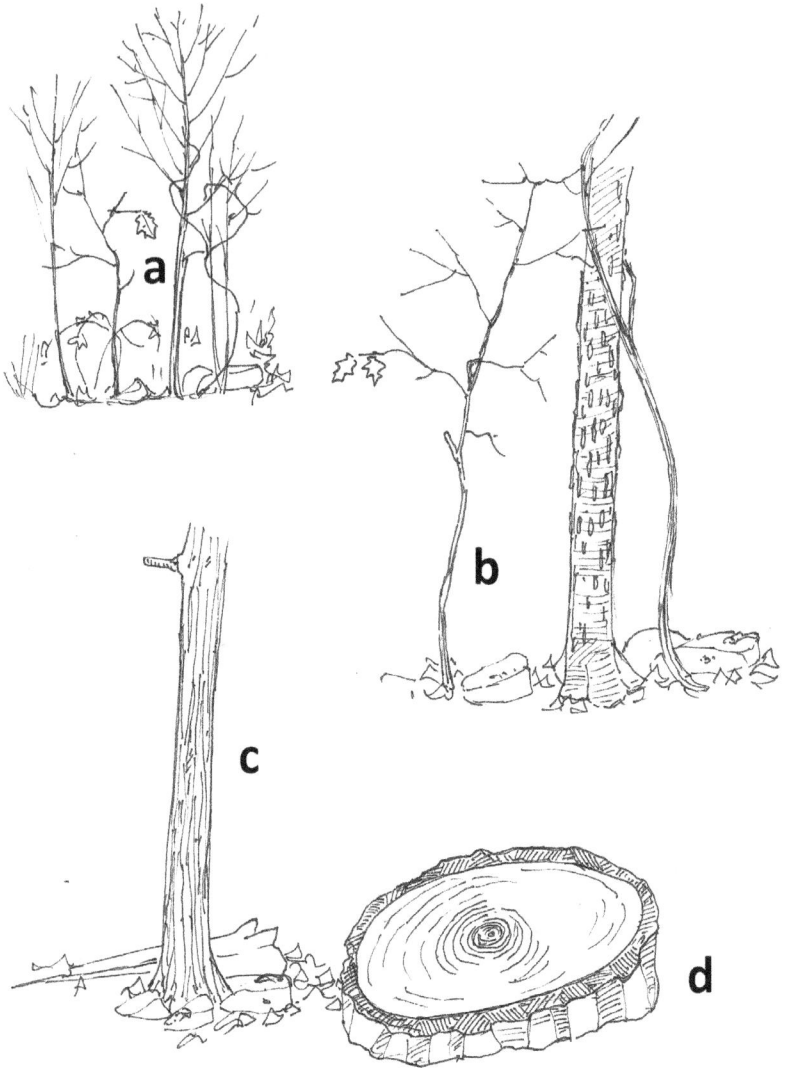

FIGURE 7.14. Mid-tolerant hardwoods such as northern red oak (**a**) often come into a disturbed forest during an early stage of ecological succession (the sapling with the retained leaf is the oak). More rapidly growing but intolerant trees such as hackberry (**b**) overtop the young oak, which is forced into suppression. The red oak eventually grows into a mature tree as the hackberry declines with old age and becomes a deadfall (**c**). This process can be seen in a cross section cut from the oak in subsequent logging: an initial period of slow growth is followed by an extended period of increasingly larger rings (**d**). Long-lived red oaks may experience several of these release pulses during their lifetime.

have figured into the life history of the trees that make up a given forest plot.

A typical example of the story that tree rings can tell in our Ozark region is illustrated in Figures 7.15 and 7.16. This example is based on a reconstruction of events on one of several study plots used to infer the status of Ozark chinquapin in our forests before the arrival of chestnut blight. All trees larger than five inches in diameter were cored on each of the study plots. The results showed that most trees on the plots today originated after logging, in the years before the land was turned over to become part of the Ozark National Forest in 1938. The few trees that originated before 1850 showed abrupt growth increases ("release"), indicating times when the forest was greatly thinned by logging in preparation for transfer of the land to what would become an outlying parcel of the Ozark National Forest. Although the forest at this site seems rather uniform on inspection, as discussed in Chapter 6, the tree-ring data show that only a few trees are more than a century old, all the others having originated (foresters say they were "recruited") in the years after the logging events identified in Figure 7.16. The recruitment of the now long-dead chinquapins represented by the fallen logs in the figure was investigated with the knowledge that they all died in 1957, when chestnut blight arrived in northwest Arkansas. The recruitment date for many of these old logs is biased toward a more recent date, since the old chinquapin wood was often hollow in the center and these extra rings could not be counted. Even so, chinquapin recruitment seems to significantly predate oak recruitment. This is consistent with the known reproductive strategy of chinquapin, which involves establishment of suppressed seedlings in the understory of closed-canopy forest with the expectation that a future disturbance will allow for rapid upward growth (see Figure 6.6). This particular example shows the detail that a professionally conducted forest forensic investigation can yield, but this kind of information about release and recruitment can often be obtained by carefully examining the exposed cross sections of the many trunks cut to clear forest trails throughout the Ozarks (as one of us is shown doing in Figure 7.11).

There are a few occasions when even casual examination of a tree's cross section can show a dramatic story written in the growth rings.

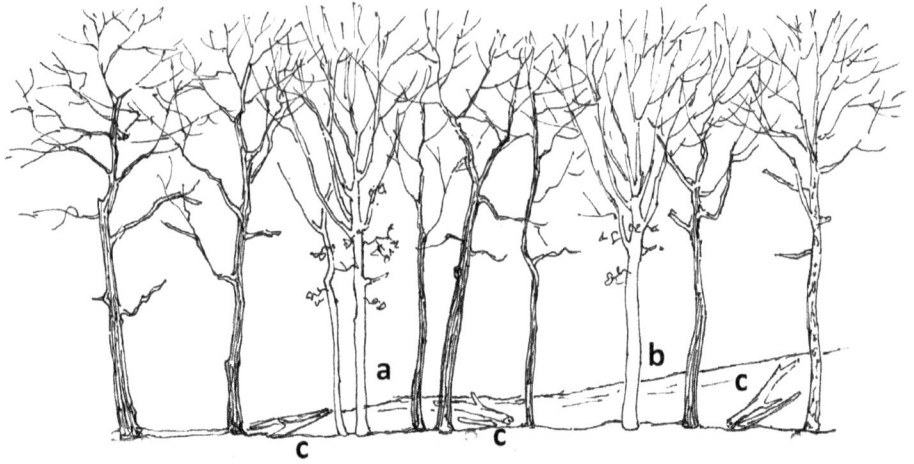

FIGURE 7.15. Schematic cross section of a test plot in a ridgetop stand (also shown in Figure 6.3) in the Ozarks where forest history was investigated by coring all trees that were more than five inches in diameter. Post oaks (**a, b**) were the oldest trees on the plot. Fallen logs (**c**) are the remains of Ozark chinquapin killed by chestnut blight. The other trees are black oaks (shaded trunks) and a white oak (mottled trunk).

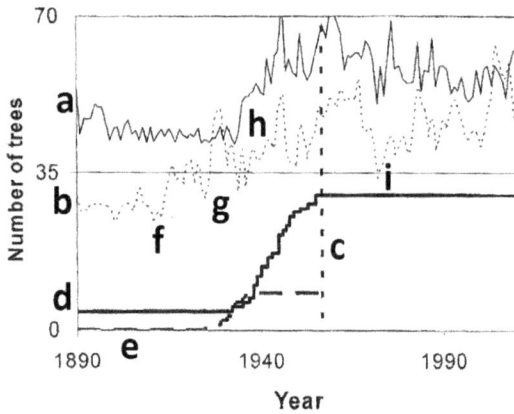

FIGURE 7.16. Tree-ring analysis of the stand represented in Figure 7.15: (**a, b**) growth-increment history of the two oldest post oaks (labeled **a** and **b**, respectively, in Figure 7.15); (**c**) arrival of chestnut blight; (**d, e**) recruitment of oaks and chinquapin; (**f–h**) abrupt increases in growth rate ("release events") in the older trees that represent episodes of forest cutting before the transfer the land to the U.S. Forest Service in 1938; and (**i**) an abrupt decline in growth rate reflecting an extreme drought in the summer of 1981.

Basal area increment per decade in square inches

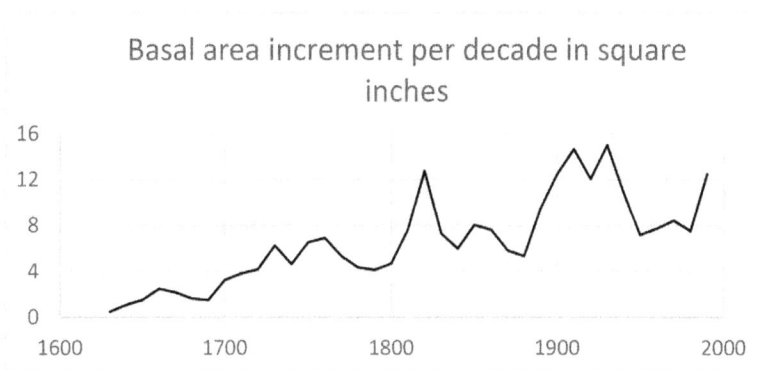

FIGURE 7.17. Profile of a 360-year-old black gum in central Massachusetts, where tree-ring increment data (plotted as a smooth curve by summing growth increments for each decade) show how major hurricanes in 1788, 1821, and 1938 reduced growth for extended periods by ripping branches from the tree's crown. Damage from the 1938 hurricane was still evident in 1990, when the core was taken.

Figure 7.17 shows an example of a core taken from a black gum in central Massachusetts. Black gum also occurs in the Ozarks, but the eastern site location has the advantage of more than four centuries of direct historical records of events. In this particular tree's rings, three periods of dramatic growth decline were apparent to the naked eye. Plotting of decade-totaled ring thickness serves to smooth out some of the year-to-year variations in the data, enabling us to see that the three major slowdown periods correspond to the dates of three major hurricanes known to have traveled up the Connecticut River Valley. Damage incurred in the famous 1938 hurricane is clearly seen in the top of a profile of the tree—making for an interesting story, even if it comes from far outside the Ozark domain.

CHAPTER 8

The Flow of Water

As the primary agent of erosion, water flow is one of the most important forces affecting the Ozark forest landscape. Water is even more directly relevant to the outdoor enthusiast because so many Ozark trails follow streams over much of their length. Flowing water is just naturally interesting in all its many forms, and even casual observers can sense that specific patterns reveal aspects of the underlying geology of the stream features they encounter. Waterfalls are always of great interest, and everyone enjoys looking into the crystal-clear water of mountain streams during periods of normal flow. There is a specific discipline, called "geomorphology," devoted to the study of how the force of water and other natural mechanisms determine the shape of the landscapes we encounter. A general introduction to the science of geomorphology can go a long way toward enhancing your enjoyment of the Ozark scenery while engendering a greater appreciation of our natural water resources.

Water is delivered to the Ozark landscape in the form of rainfall. When the rainfall rate exceeds the capacity of the land's surface to absorb it, the excess begins to flow off downhill. This starts out as "sheetflow"—simple downhill flow along the surface of the ground. After traveling some distance, the sheetflow encounters a local depression or other feature that collects sheetflow and concentrates it, such that there is enough erosive power to excavate a channel. This happens in episodes, but there will be some distance downhill where the processes that tend to flatten such temporary channels cannot erase them between rainstorms, and the head of a natural stream channel is formed (Figure 8.1). Downgradient from this point, the stream channel serves as the primary conduit for erosion products delivered from the slopes above.

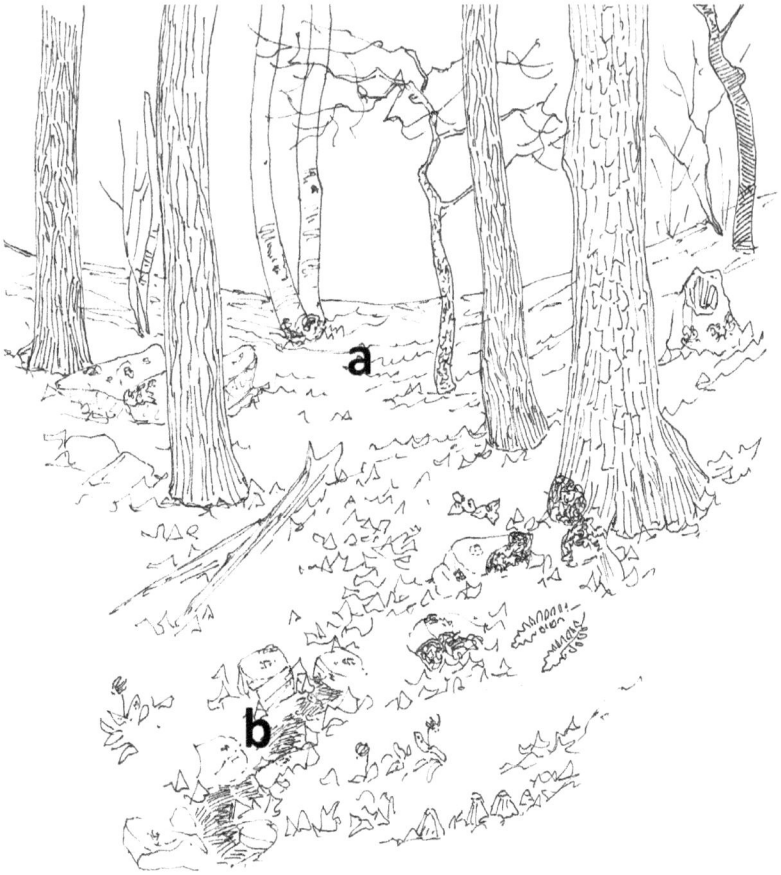

FIGURE 8.1. Small headwater hollows (**a**) are the locations where overland sheetflow becomes concentrated into discharges with enough erosive force (**b**) to begin to define a first-order channel at the head of an Ozark drainage network.

Streams and rivers carry sediments from their source area in three forms. These are (1) as dissolved solids permanently incorporated into the water, (2) as suspended particles of sand and clay held in the water by the natural turbulence of flow, and (3) as large particles ("bedload") that are literally pushed downstream by water flowing above. The lovely turquoise green we associate with many of our Ozark streams, such as the Buffalo and the Jacks Fork, is provided by dissolved minerals in the water and is typical of rivers that drain land with lots of exposed limestone or dolomite. Suspended sediment is what gives such rivers that familiar muddy brown during periods of unusually high water. The bedload of gravel and cobbles remains largely in storage along the stream course and is mobilized only during periods of extremely high water.

The large amount of sediment effectively in storage along a stream course forms much of what we perceive as the stream habitat. The stream itself is seen to flow through a well-defined channel (Figure 8.2), the uniformity of which is surprising when you think about it. Hydrologists who have studied this phenomenon have identified the importance of "bankfull flow" in determining channel size and structure. Bankfull flow is the amount of water flow that just fills a channel of given width and depth without overflowing onto the surrounding banks. On average, bankfull flow occurs about once a year. Lower flows occur more frequently but don't have enough energy to do much work. By contrast, unusually large and catastrophic flows can do quite a bit of damage, but while they can permanently change the shape of the stream valley, they don't happen often enough to permanently change the geometry of the channel that conducts water flow within that valley. Bankfull flow provides just the right combination of frequency and erosive energy to determine the size and shape of the streams we encounter on our excursions.

One result of a stream or small river having to flow through its bedload is the presence of pools and riffles (Figure 8.3). The movement of gravel and small cobbles during storm events causes natural "traffic jams" in the rocks being pushed downstream by these higher flows. The result is that you see riffles where water cascades over and around the piles of gravel and then flows into quiet pools in the regions between these sections of faster flow. Hikers will return to sections of trail along their favorite streams and see pools and riffles in the same places every year. Fishermen will return year after year to the same hole that always seems to hold a feisty smallmouth. But this landscape is not quite as static as it might look. Hydrologists have studied riffles by painting rock samples at low water. After storms the riffles look the same, but the painted rocks are found in other riffles downstream. Just as in traffic jams that seem to develop at predictable locations on the highway, it's the same old traffic jam but a new set of vehicles every rush hour.

We should note that many of the studies describing these processes

FIGURE 8.2. Streams flow through well-defined channels that have properties related to the erosive power of the water flowing through them; the channel width (distance between **A** and **B**, about four feet, in this example) is determined by the average yearly maximum or "bankfull" discharge.

were conducted in stream environments where water flows over relatively thick bedload deposits. Ozark streams are actively eroding their way into consolidated bedrock, and this hard surface cannot be deformed by flowing water except through the excruciatingly slow process of abrasion by entrained sand and gravel. So the ability of our rivers to interact with their sediments is often limited, as illustrated in Figure 8.3 by the exposed ledge that extends across the center of the pool, an extreme case of erosional restriction that results in waterfalls. Some rock layers within the Ozarks are much harder and more resis-

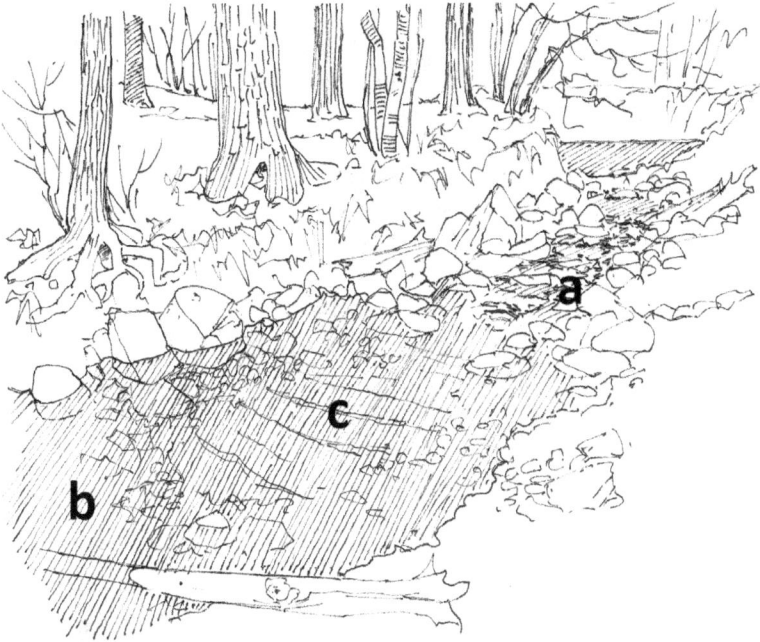

FIGURE 8.3. Under most flow conditions, streams do not carry enough flow to push their bedload of gravel downstream, and pulses of gravel accumulation form sequences of riffles (**a**) and pools (**b**). Note that Ozark drainage systems are actively cutting into the Ozark uplift today, so bedrock (**c**) is never far below the base of stream channels and is often exposed in the beds of small rivers like the Meramec and Buffalo.

tant to abrasion than many others, and water is forced to cascade over them (Figure 8.4). The cascade can be in the form of a direct plunge, where a single thin and hard bed defines the waterfall. In that case, the water and sediment mixture falling off the top of this ledge excavates a plunge pool below, and the force of the falling water erodes the softer rocks beneath the more resistant ledge above. An ever-enlarging cavity eventually cannot support the weight of the overlying ledge, causing it to break off, and the process begins anew as the waterfall slowly retreats upstream. In other examples, the waterfall occurs as a series of cascades down a steep slope of uniformly hard rock. Both of these situations are depicted in Figure 8.4.

The bulk of sediments being carried by streams is commonly

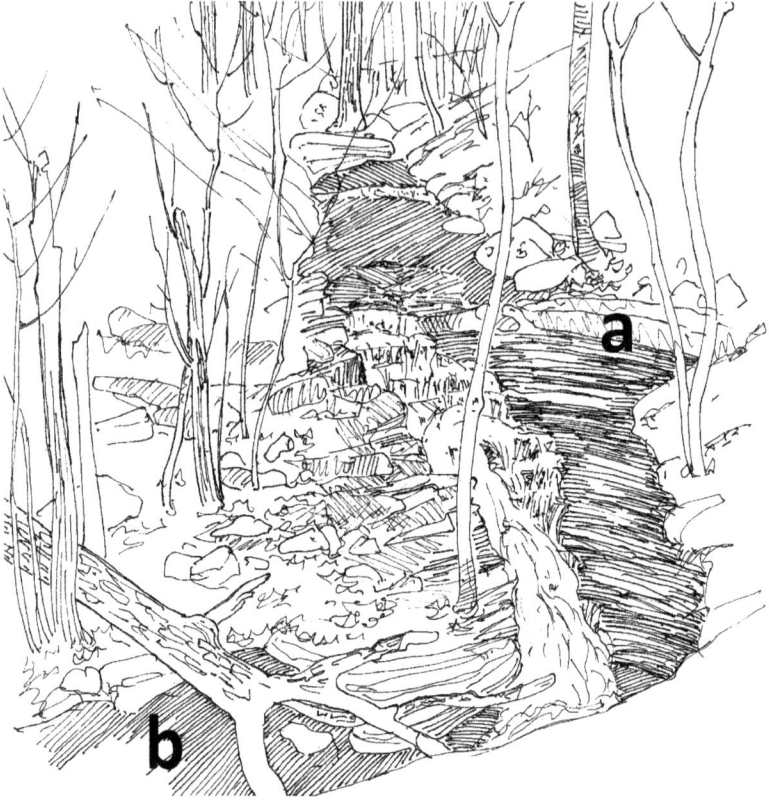

FIGURE 8.4. Two examples, *above and facing page*, showing how streams develop waterfalls where they encounter especially hard layers in the rocks they are cutting into (**a**) and how the force of the water plunging over the falls often excavates a plunge pool (**b**) in the softer rocks at the base of the falls.

stored in terraces that develop on either side of the stream channel that carries the normal flow (Figure 8.5). These terraces form relatively rich and moist habitats with soils that can be especially fertile. Terrace sediments are deposited during high flows when fast-moving water in the main channel spreads out onto the terrace surface and suspended sediment settles out in the slower water. This accounts for the fine-grained and clay-rich terrace sediments responsible for soil fertility. The stream channel itself tends to wander back and forth laterally within the terrace environment, a process known as "mean-

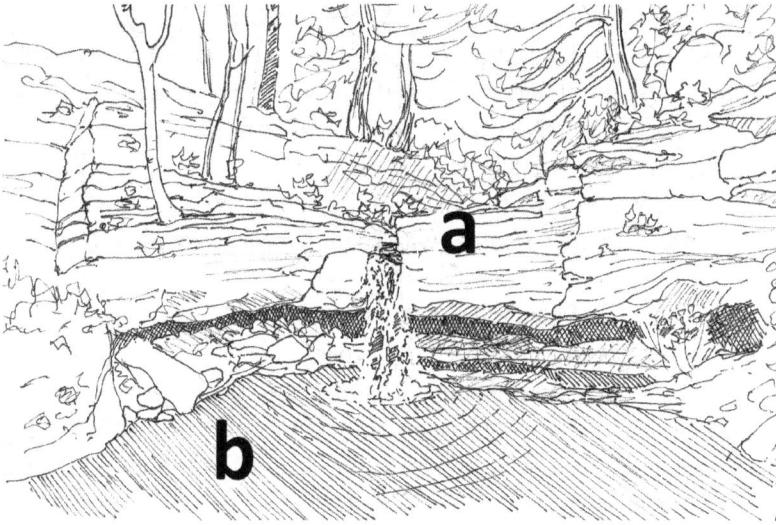

dering." The transport energy of water in stream channels is indicated by the slope of the streambed. During the bankfull flows that are most effective in determining channel structure, the stream may have more erosive energy than needed to carry the sediments it is transporting within the channel. Meandering results as a way for the stream to adjust its sediment-carrying capacity. A meandering stream effectively increases its length as a way to reduce its gradient and equilibrate itself to the amount of sediment being carried. For example, if a river wanders a distance of two miles in the process of going a mile downstream, it has effectively reduced the slope of its bed by half, thereby reducing its gravitational work capacity by the same amount. But there is a second interesting sediment-transport process going on here. Once a stream starts to go around a meander bend, it generates a centrifugal force that has to be opposed to get the water through its turn. This opposition force is generated when water piles up against the eroding bank. This force turns the flow to the side to conduct water around the bend, but it also enhances the local eroding power of the stream. That's because the water near the bottom is not flowing as fast and thus experiences much less centrifugal force. So the high pressure from above effectively pumps sediment toward the inside of the bend along the bottom of the stream. This process keeps the full force of the water applied to the eroding bank while piling up sediment on the inside of the bend, a feature hydrologists call a "point bar." Looking down at these point bars from a cliff, one can often see arcs of vegetation getting progressively larger and older (a progression from herbs to seedlings to saplings), which shows how the sediment pumping has progressively expanded the inside of the meander. Canoe campers, of course, are intimately familiar with point bars as convenient places to spend the night.

The process of meandering is easy to understand in principle but is seriously complicated in the Ozarks because we have what are effectively meanders within meanders. That is, the stream channel meanders back and forth within its contemporary floodplain, but the valley itself has been eroded deeply into solid bedrock over eons (Figure 8.6). These bedrock meanders must have been generated at some dis-

FIGURE 8.5. Larger streams develop small terraces (**a**) where sediment is deposited as floodwater from a major storm slows when it overtops the central stream channel. In the long periods between such extreme events, much more frequent but less intense high-water flows cause the stream channel to wander laterally, cutting into these terrace deposits (**b**). The violence of high-water events is evident in tree roots exposed by scour (**c**) and by the clumps of debris suspended in lower branches (**d**).

tant time when the Ozark Plateau was first subjected to its uplift, and meanders that were initially generated in soft surface sediments were superimposed on bedrock as the early streams cut their way down. Hydrologists have generated formulas that relate the size of a river (discharge and channel width) to the wavelength of the meander. It is known, for example, that meandering streams typically have a wavelength equal to seven channel widths. The difference between valley meander size and the meanders in the streams they now contain can be used to show that the amount of water carried by rivers in these valleys has changed significantly over time. An extreme example is the identification of large-scale valley meanders in western Tennessee that are attributed to an ancestor of the modern Mississippi that may have had many times the flow of today's already great river.

One byproduct of the stream and river environment is the periodic effect of major and occasionally catastrophic floods that can inflict great damage to streamside trees. As a result, many of the trees that grow in this environment are adapted to deal with broken trunks and exposed roots (Figure 8.7). The damage and erosion incidentally adds interest and intriguing detail to the scenery we encounter on Ozark hikes. One commonly seen example is a great contorted mass of tree roots exposed in an eroding bank (Figure 8.8). Such scenes are interesting to look at and provide insight into root structure that normally remains hidden from view. Yet another specific component of Ozark streams is the canebrake (Figure 8.9). Big cane (or "river cane") is a North American bamboo that can form dense thickets on the fertile soils of river terraces. Cane is probably less common today than it was in prehistoric forests, because fertile stream and river bottoms have largely been converted to cropland and pasture. Memoirs from the earliest European settlers in America—such as Henry Timberlake, held hostage by Cherokees in the mid-eighteenth century—remind us how useful those dense canebrakes were when it came time to hide from hostile war parties.

The karst nature of many locations in the Ozarks accounts for the numerous springs that are often seen feeding stream channels when water is low (Figure 8.10). These springs most often have small discharges and appear as water seeping out from bedding planes within or at the base of limestone strata. Water infiltrating downward through

FIGURE 8.6. Rivers tend to generate meanders by lateral movement of channels in the sediments that form their banks. Meander lengths are about seven times the width of the river channel. Ozark rivers meander within channels deeply incised into uplifted bedrock, the size of the meanders representing the properties of the ancestral drainage system that originally started to cut into the uplifting landscape.

fissures in the rock encounters these horizontal planes or the base of the brittle and fractured limestone rocks and flows laterally, to emerge where a stream channel has cut down into the bedrock. The flow from these small springs often extends only a short distance downstream and then sinks into another set of fissures in the rock (Figure 8.11). Pioneers would be careful to note the presence of minnows in these small sections of flowing and sinking streams, an indication that the

FIGURE 8.7. Tree species such as sweetgum, river birch, and the giant sycamore shown here (dwarfing a resting hiker) are adapted to survive and withstand battering by flood debris during high-water events by resprouting from battered trunks and broken lower branches.

FIGURE 8.8. Stream erosion in high-water events often undercuts the trees lining stream banks, creating intricate exposures of roots (or "root plates") that can expose the belowground anatomy of trees that would otherwise remain hidden from view. When these root plates tumble into the stream channel, they become hazards referred to by boaters as "root wads."

spring did not dry up during periodic drought. Hikers, too, find it interesting to look for such effectively isolated populations of small minnows. We have been surprised, when revisiting such locations after major flood events that produced drastic changes to the stream environment, to find that the little minnows seemed to be just about as numerous as they were before the flood. How did such delicate little

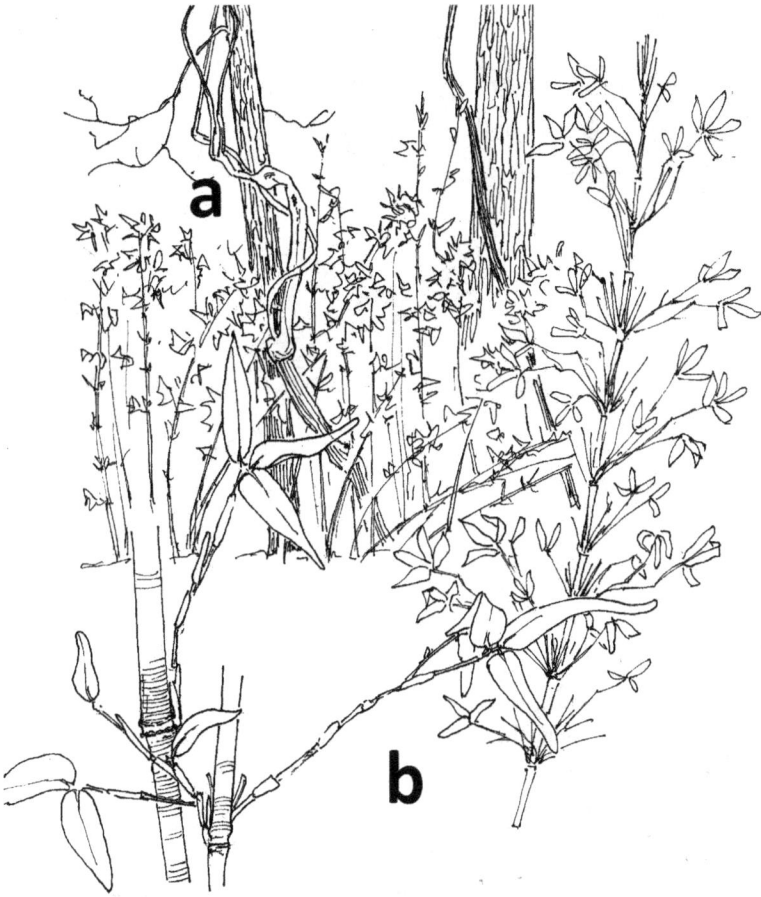

FIGURE 8.9. An unusual feature of river banks in the southern Ozarks is the presence of a form of bamboo known as "river cane," which early settlers often found to be extensive in fertile bottomlands that have long since been converted to plowed fields or pasture. The cane thickets (**a**) are often tall and thick enough to have hidden early pioneers from hostile Native Americans. Close inspection of cane stalks (**b**) shows the hollow tubes and joints we identify with bamboo.

creatures manage to survive such devastation? The vast majority of Ozark springs are relatively small, but a few large springs have become important tourist attractions (Figure 8.12).

Another interesting feature of small Ozark mountain streams is the result of an event hydrologists call a "channel evulsion." This occurs in extreme storm events when a temporary obstruction, such as a debris dam, forces the stream channel to make a new course, abandoning

FIGURE 8.10. Ozark streams often receive inflows from small springs embedded in the bedrock exposed along the stream channel; here, water flows out from a bedding plane in a chert layer intersecting an otherwise dry streambed.

the well-defined usual channel. Afterward, the obstruction may be removed and the original flow path resumed or the stream may be permanently diverted. Either way, a secondary channel is created where there will be little flow most of the time, and several quiet, clear pools will remain (Figure 8.13). These little secondary pools can be havens for amphibians and other life-forms. In the process, they can provide an interesting photographic opportunity: snails grazing on the algal films that grow on boulders in these pools generate intricate patterns of intersecting pathways that show where they have been feeding.

One additional class of water body does not involve flow at all. "Vernal pools" form where water fills shallow depressions during the wet spring months, only to dry out later in the year (Figure 8.14). In the Ozarks, these pools naturally occur where the hummocky terrain produced by a landslide (see Chapter 2) has created numerous small depressions. In modern Ozark settings, these temporary pools

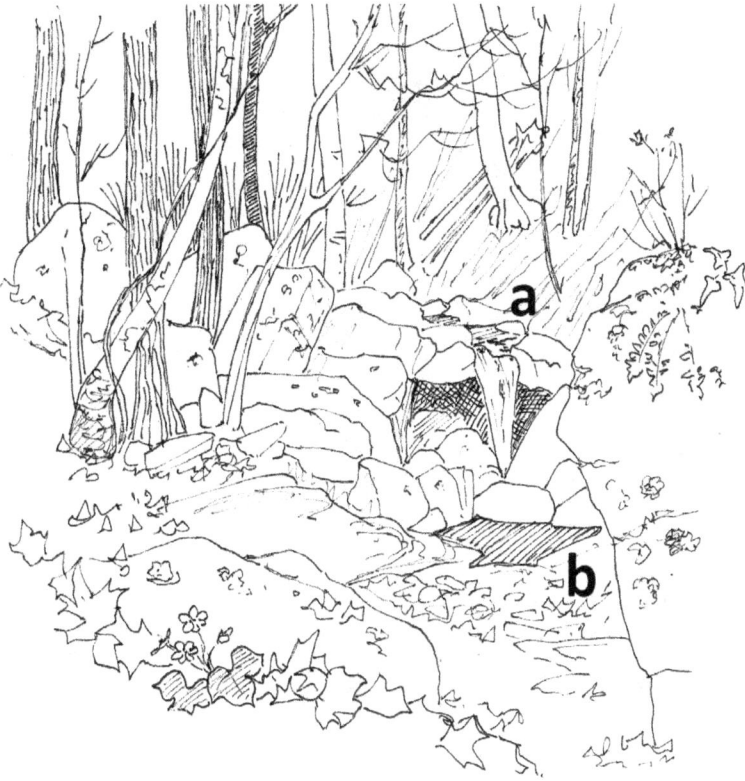

FIGURE 8.11. Many Ozark streams whose flows arise at springs will have "losing" or sinking sections where the flow in the streambed (**a**) disappears into the bed of the stream some distance downstream (**b**).

are more likely to be caused by wheel ruts in old forest roads or in ditches along established roadways. Such depressions trap leaf litter and other debris, which become rich food sources for the larval stages of amphibians and insects. The fact that such pools are not well connected to perennial waterways and dry out later in the year means that fish predators are not present. To see signs of amphibian activity, look for gelatinous masses speckled with black eggs or swarms of little tadpoles. On the first warming days of early spring, the clacking of wood frogs as they mate in these pools can sound like flamenco dancers going wild with their castanets. In our woods, this is often the first definite sound declaring that spring is really just around the corner.

Although once nearly exterminated across the southern part

FIGURE 8.12. Most Ozark springs are modest and may not flow at all during dry periods, but a few truly impressive springs, such as Blanchard Springs (shown here) and Roaring River, are major tourist attractions.

of North America, beavers have more recently begun to reappear throughout much of their former range, including the Ozarks. You can see beaver ponds at many locations in our area, along with other indications that beavers are present. These signs include the starkly white color of freshly peeled branches lying in shallow water and the sharply pointed stubs of small trees and shrubs where these woody-stemmed plants have been chewed off by beavers. Beaver dams are constructed when branches and twigs are pushed into the stream and collect against natural obstructions in the flow (Figure 8.15). These sites trap additional debris, and beavers use them as the basis for their deliberate construction of a dam. The resultant flooding of the stream channel—and the beavers' removal of shade-producing trees such as sycamore and elm, which have the most palatable bark—allows a lush

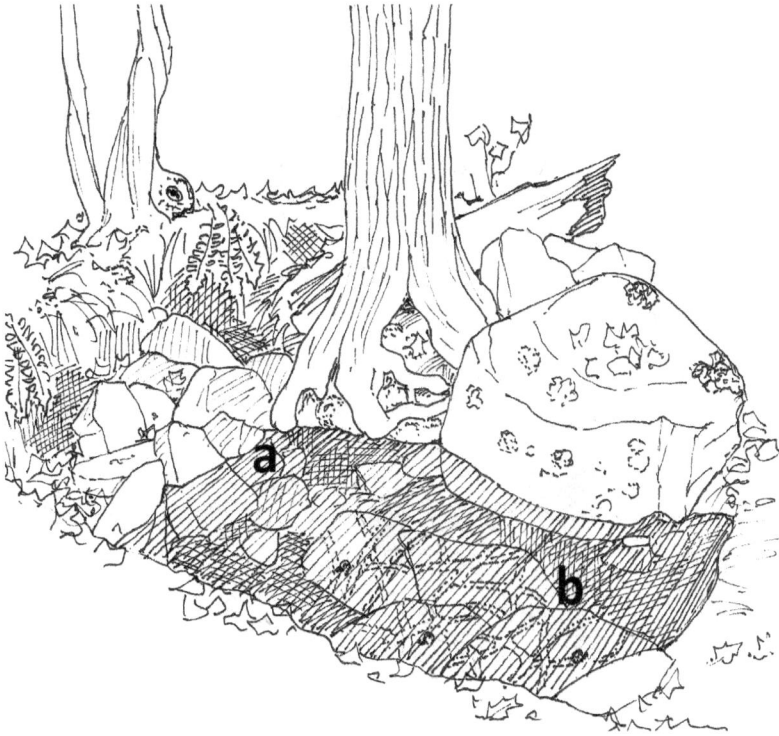

FIGURE 8.13. One interesting type of habitat along Ozark streams is created by a process known as "stream evulsion," in which a major storm causes the stream channel to move laterally, abandoning its former bed. Clear and quiet pools remain in the abandoned channel (**a**) to form fertile habitat for aquatic life where snails crawling across algae-encrusted boulders leave intricate pathways of clean rock (**b**) as if designed specifically for the amusement of hikers.

herbaceous meadow to develop around the pond. But the presence of a beaver lodge in the center of the pond is not a common sight. Most of the time, the beavers simply take shelter under an overhanging bank, or burrow into the terrace sediments to become "bank beavers." They are also hard to see because most of their activity is nocturnal. Given that a mature beaver is as large as a medium-sized dog, they have relatively little fear of predators in the Ozarks and are likely to increase in numbers. So we can expect to see more of them in the future, along with the damage they inflict—as in the case of the rogue beaver now gnawing on landscape trees at the Crystal Bridges Museum of American Art in Bentonville, Arkansas.

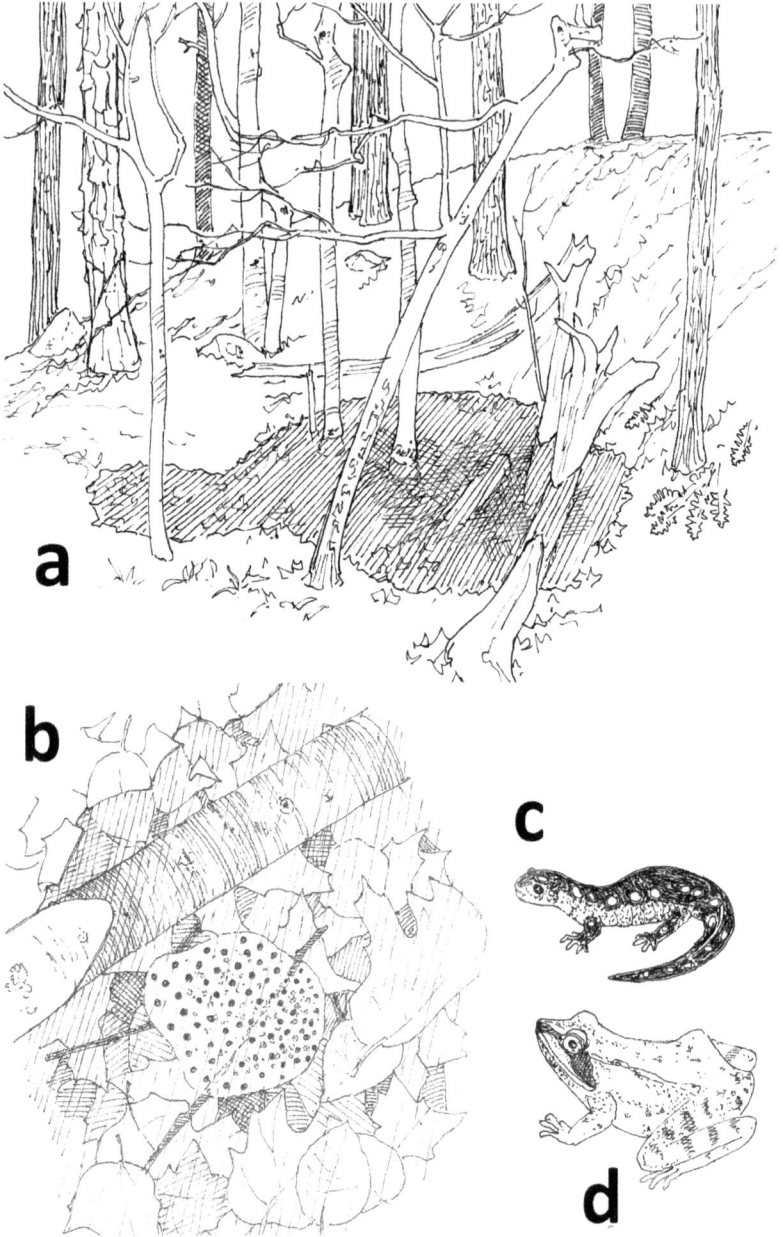

FIGURE 8.14. An ecologically important water formation in Ozark forests, the vernal pool (**a**), is created when a hollow (for example, in the hummocky soil of landslide deposits) becomes a small, temporary pool in the wet spring season. The decaying forest litter (**b**) in such pools provides a vital source of nourishment for amphibians such as the spotted salamander (**c**) and wood frog (**d**), which lay their eggs in these pools.

FIGURE 8.15. Beavers begin to dam a stream by simply allowing debris, including starkly white branches from which they have eaten all the bark, to collect and begin forming a natural barrier in the channel (**a**), which becomes a dam (**b**) that creates a pond surrounded by lush grasses and herbs (**c**). Note the pointed stumps of small trees that the beavers have cut for food (**d**), meanwhile creating pathways where tree branches are carried into the pond from more distant food trees (**e**). The rising water level drowns some larger trees (**f**).

FIGURE 8.16. A small Kentucky spotted bass is attracted to the minnow-like lure protruding from the shell of a *Lampsilis* mussel embedded in the gravel of an Ozark stream. If the bass attempts to seize the imitation minnow, it will be sprayed in the mouth by a stream of mussel larvae that can attach themselves to the fish's gills, where they can obtain nourishment from microscopic food items passing through the gill chambers while hitching a ride to a new permanent residence.

One other aspect of flowing water in the Ozarks is the presence of freshwater mussels in the riffles of small rivers and larger streams. Hikers often think of these as "clams" because they resemble the edible bivalves consumed at seafood restaurants. You will see their shells deposited on gravel bars and can readily find living mussels embedded in the streambed if you look for them. For much of their lives, freshwater mussels are stationary filter feeders that live on organic debris ingested from the stream water passing over them. You might think of them as just fixed features of the stream bottom that are hardly different from the pebbles and stone cobbles around them. However, there is actually a wide range of different species of mussels, each of which has an intimate relation with a specific variety of fish needed to complete its own life cycle (Figure 8.16). For example, the *Lampsilis* mussel of Ozark streams has a specially designed reproductive organ that is configured to spray mussel larvae into the mouths of fish that are enticed to approach for just that purpose. The mussel larvae attach themselves to the gills of the host fish, where they mature into tiny mussels that can drop off at a convenient location to complete life as a stationary filter feeder. Some species of mussels are thus closely tied to their fish partners, and a change in the ecosystem can push them to the verge of extinction. This has happened in East Asia, where the loss of salmon runs from developed watersheds has threatened or eradicated some native mussel species. In the Ozarks, one threat to mussels is posed by the conversion of former warm-water reaches of rivers to cold tailwater environments below major river impoundments. Warm-water species of fish such as catfish and drum no longer ascend tributaries from these altered habitats, incidentally breaking the reproductive cycle of species of mussels that require their presence in headwater streams. The increased load of sediments in other Ozark streams has also had an adverse effect on mussels. There is little doubt that a century ago, mussels as a group were more abundant and diverse in streams throughout Arkansas and adjacent states. Conservation of threatened and endangered mussel species is thus a major consideration in the management of land in our region by the National Park Service and other federal and state agencies.

CHAPTER 9

Shrubs, Vines, and Understory Trees

In a well-developed Ozark forest dominated by broadleaf trees, it is usually possible to recognize seven distinct strata of vegetation. These are the canopy (or "overstory"), subcanopy ("understory"), sapling, shrub–vine, seedling, herbaceous, and ground layers. The canopy layer is made up of larger trees, such as various oaks and hickories that are usually more than fifty feet tall (sometimes exceeding seventy-five feet on favorable sites). The subcanopy consists of smaller trees (typically no more than twenty to thirty feet tall), including those that have the potential to grow taller (thus possibly reaching the canopy at some point) as well as species that are incapable of growing very tall. Prominent examples of small trees that only rarely achieve canopy size are dogwood, redbud, serviceberry, and hophornbeam (also known as "ironwood").

Dogwood (or "flowering dogwood") is one of the most familiar of all trees, large or small (Figure 9.1). Throughout much of eastern North America, the appearance of dogwood flowers is almost universally recognized as a sure sign that spring has arrived. However, what most people regard as a single large "flower" is actually a composite structure consisting of a tight cluster of small and relatively inconspicuous yellow-green flowers surrounded by four very conspicuous, nearly inch-long white "petals." These "petals" are actually bracts (modified leaves) and not true flower parts. Later in the year, dogwoods produce somewhat elongated, bright red, berry-like fruits called "drupes" that are about half an inch in length. Although poisonous to humans, these shiny red fruits are consumed by birds. An early frost can sometimes cause dogwood berries to ferment, resulting in some strange avian behavior when flocks of migrating robins become tipsy from overconsumption.

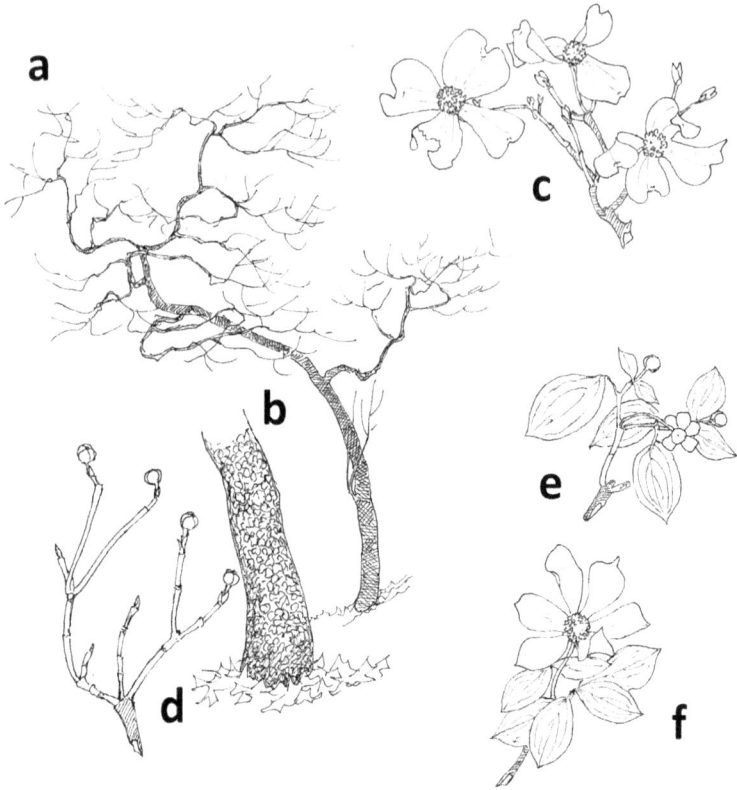

FIGURE 9.1. A familiar small tree found in the understory of Ozark oak–hickory forests, dogwood (**a**) has a distinctive light-brown bark broken into fine flakes (**b**). White and occasionally pale pink flowers appear in the early spring, before trees have leafed out and when the dogwood's own leaf buds are just beginning to expand (**c**). The flower buds form at the end of the growing season, so they are apparent on branch tips through the winter months (**d**), and the flowers give way to clusters of bright red berry-like drupes in early fall (**e**). Some other species, such as the Pacific coast dogwood (**f**), produce similar large floral displays, but our flowering dogwood enhances its display by producing flowers before they are obscured by the new leaves of the year.

Among the many species of dogwood found in the Northern Hemisphere, the flowering dogwood we have in the Ozarks is unique in that the flowers are produced before the leaves, whereas other dogwood species that have large flowers (like the North American Pacific dogwood) bloom when the crown of the tree is fully leafed out.

Having flowers embedded in a mass of foliage obviously detracts from the floral display. Our dogwood produces what looks like a compact flock of white butterflies in the somber springtime woods when buds are just beginning to swell, and hikers appreciate the relief from the drab colors of late winter. The dogwood's flower buds are produced during the latter part of the growing season and are conspicuous on branch tips after leaves are shed, so that hikers can get a preview of locations where the dogwood floral display is scheduled to be especially intense and plan future outings accordingly.

Even without the flowers, dogwoods are easy to recognize. First, the leaf arrangement is opposite, a condition that is much less common than the alternate leaf arrangement characteristic of most other trees found in the same forests as dogwood. Moreover, the lateral veins of the leaf are distinctive in that they curve upward and away from the leaf margin. In addition, the bark is dark reddish-brown and rough, with many small, square-like plates. The wood of dogwood is very hard and has been used to make a wide variety of things, including tool handles, wooden gears, and the heads of mallets and golf clubs. The tree often grows in a form that appears layered and laterally spreads out as the branches expand to seek shafts of sunlight that penetrate the canopy above.

There are several other distinctive, spring-flowering subcanopy trees in Ozark oak–hickory forests (Figure 9.2). The flowers of redbud (Figure 9.3, b) appear at about the same time as those of dogwood. In spring it is difficult not to notice these small trees, their bare branches covered with dense clusters of showy pink flowers. Closer inspection of an individual redbud often reveals that some of the flowers arise directly from the trunk, a most unusual feature for trees in temperate forests but not uncommon in the tropics. The numerous small flowers resemble, except for color, those of the common garden pea. This similarity is due to the fact that both plants belong to the pea family (Fabaceae). As in the case of dogwood, Ozark redbud flowers appear ahead of the leaves, which is not the case for other redbud species of the American West or the Mediterranean region (the latter is more commonly known as the "Judas tree"). The leaves of redbud are distinctly heart-shaped and relatively large (often as much as five inches long and wide). The fruits are flattened, pea-like pods that are two to

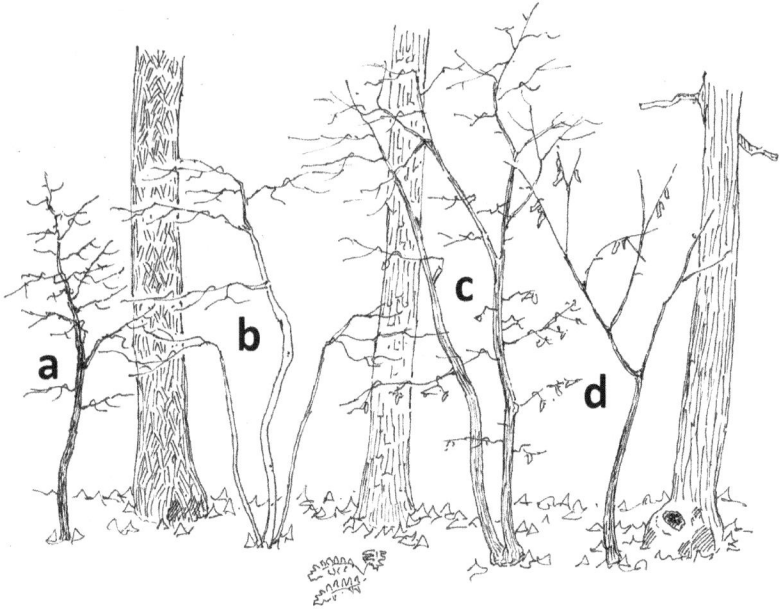

FIGURE 9.2. Typical midwinter scene in the Ozarks, showing the stature of common subcanopy trees in an oak–hickory forest: (**a**) Mexican plum, with nearly black bark and many short fruiting "spurs" on the branches; (**b**) serviceberry, with silvery bark and outward-leaning layers of fine branches; (**c**) ironwood, with brown fibrous bark and dried-up leaves still attached to lower branches; and (**d**) redbud, with dark gray bark and a few ripened seed pods still clinging to upward-arching branches.

four inches in length. These remain on the branches after the leaves are lost in the fall, making this an easy tree to identify even during winter. Native Americans used redbud flowers as food, but today the tree's most common use is as an ornamental in yards and parks. The same is true of dogwood, perhaps the most frequently planted ornamental tree in much of eastern North America.

Serviceberry (also called "shadbush" or "Juneberry"; Figure 9.3, a) is the first tree to produce conspicuous flowers in the forests of the Ozarks each spring. The flowers have five strap-shaped white petals and occur in loose clusters on branches that are still devoid of leaves, unlike other North American species of serviceberry in which drab white flowers are embedded in massed foliage. In our species, the expanding leaves are present before leaf-out and have a distinct ruddy color that gives

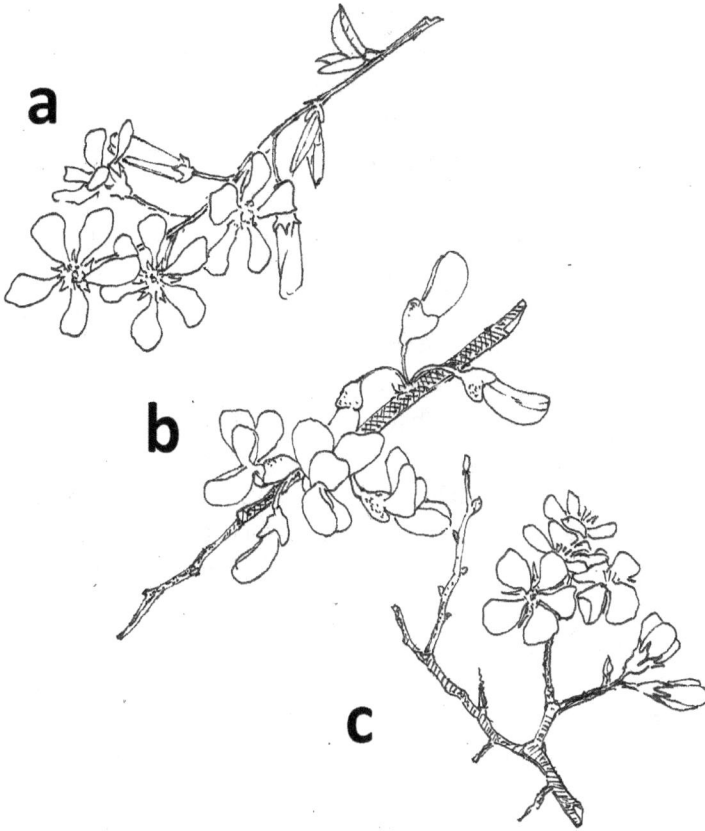

FIGURE 9.3. The showy flowers of (**a**) serviceberry, (**b**) redbud, and (**c**) Mexican plum.

the dense white cloud of serviceberry flowers an attractive pinkish tint. Although generally considered a small tree, serviceberry sometimes grows tall enough (occasionally more than sixty feet) to reach the canopy. There are several possible explanations of how this tree received its common name. One of the more widely known is related to the use of the flowers in church services (especially weddings and funerals) in the Appalachian Mountains where, because of harsh winter conditions, traveling preachers were unable to visit their churches until the return of favorable weather in early spring. This time of the year just happened to be the flowering season for serviceberry, which provided the only dependable source of flowers for use in church services—hence the common name. As for the tree's other common names, "Juneberry" is derived from the fact that the fruits begin to ripen during that month, while "shadbush" refers to a type of fish (shad) once found in some numbers in larger rivers throughout eastern North America. These fish spawned at the same time of year that the trees were in bloom. The fruits are small (usually no more than about half an inch in diameter) and apple-like (which is not surprising, since serviceberry and apple belong to the same plant family, the Rosaceae). They are eaten by many animals and were once an important food item for Native Americans. Children must have been stationed to ward off fruit predators, because serviceberry fruits in suburban parks in our region are completely consumed by birds before they are even fully ripe. Although rarely collected and consumed by humans today, the fruit of serviceberry is actually as good to eat as a blackberry or blueberry. Some other, more shrub-like forms of serviceberry are now being sold for commercial cultivation to produce abundant crops of purple-red fruit marketed as "Saskatoon berries."

Yet another, somewhat less common, spring-flowering subcanopy tree is the Mexican plum (Figure 9.3, c). There are many varieties of plum growing in North America, several of which can be found in the Ozarks. But Mexican plum is another of those flowering shrubs that produce flowers in early spring before they can be obscured by emerging foliage. The flowers are pure white and larger than those of other plum species, being about the same size as apple blossoms. Most hikers would simply confuse these white blossoms with those of serviceberry, which are present at about the same time. But this distinctive little plum should be appreciated in its own right.

Musclewood (or "hornbeam") is a characteristic small tree found along streams throughout the Ozarks (Figure 9.4). The most distinctive feature of musclewood (and the basis for its common name) is that in cross section, the trunk and larger branches are not round but are somewhat irregular in shape and have longitudinal ridges that (with a little imagination) are suggestive of the muscles of an arm or leg. The leaves are straight-veined and have a toothed margin. This little tree is most often seen growing with two or three trunks, leaning out over small streambeds, its roots exposed by flood-scouring from frequent storm events.

FIGURE 9.4. Foliage, seed cluster, lower trunk, and profile of a typical subcanopy hornbeam or "musclewood" tree.

The leaves of ironwood are similar to those of musclewood, but the two trees can be distinguished rather easily on the basis of their bark. Musclewood has smooth gray bark, but in ironwood (Figure 9.2, c) the bark is light brown and broken up into small scales in the shape of elongated rectangles. These scales are often seen to be flaking off, giving the tree's bark a fibrous look. Moreover, ironwood tends to occur in more open forests and on drier sites, whereas musclewood is often found on the banks of streams with its roots literally in the streambed. The fruits (nuts) of ironwood are produced in a papery, cone-like structure that resembles the hops used in making beer (which accounts in part for the alternative common name "hophornbeam"). Pollen studies used to characterize tree communities during glacial and early postglacial times (Chapter 2) indicate that ironwood was a much more important part of the forest as the glaciers were receding, when the climate was cooler and drier than at present, so it was probably even more abundant in the Ozarks during those times. It is also interesting to note that both musclewood and ironwood are present in European forests, where ironwood grows as hardly more than a shrub while musclewood becomes a massive, canopy-dominant forest tree.

Anyone who has hiked in both the Appalachians and the Ozarks is surely aware that various types of shrubs are much more common in eastern forests than they are here. The primary reason is that the soils of the Appalachians are relatively more acidic than those of the Ozarks, and members of the family Ericaceae, the major group of shrubs found in eastern forests, tend to be associated with fairly acidic soils. Among the more widely distributed and abundant shrubs in that family are species in the genus *Vaccinium*, which includes such familiar plants as the blueberry, huckleberry, and cranberry. Only three *Vaccinium* species are relatively widespread throughout our region (Figure 9.5), growing in Ozark mountain soils developed from relatively acidic substrates such as sandstone or deep chert residue left by the dissolution of some limestone formations.

By definition, a shrub is a small, woody plant with multiple stems arising from the ground, but the distinction between a small tree and a large shrub is not always clear. In fact, in different regions of North America, the same species can be regarded as a shrub in one place and a small tree in another. This is certainly the case for farkleberry (aka "sparkleberry" or "tree huckleberry"; Figure 9.5, a), which is capable of reaching a height of thirty feet but is usually much smaller. The pendant, bell-shaped flowers of farkleberry appear in early summer, after the forest canopy has closed. The fruits produced later in the summer are shiny black berries about a quarter of an inch in diameter. They are consumed by birds but are not considered edible for humans, who find the berries very seedy, with just a faint apple-like flavor. Farkleberry leaves and fruits often remain on branches into the early winter, in contrast to our other blueberry species.

On dry rocky ridgetops throughout the Ozarks, the low-bush huckleberry, a close relative of the farkleberry, often forms dense populations (Figure 9.5, c). Low-bush huckleberry, as the name implies, is a much smaller plant than farkleberry, rarely reaching a height of more than about two feet. However, the fruits produced by low-bush huckleberry are highly edible and are often collected in large quantities where the plant is relatively common. (While each of these shrubs is sometimes referred to as a "huckleberry," the true huckleberry

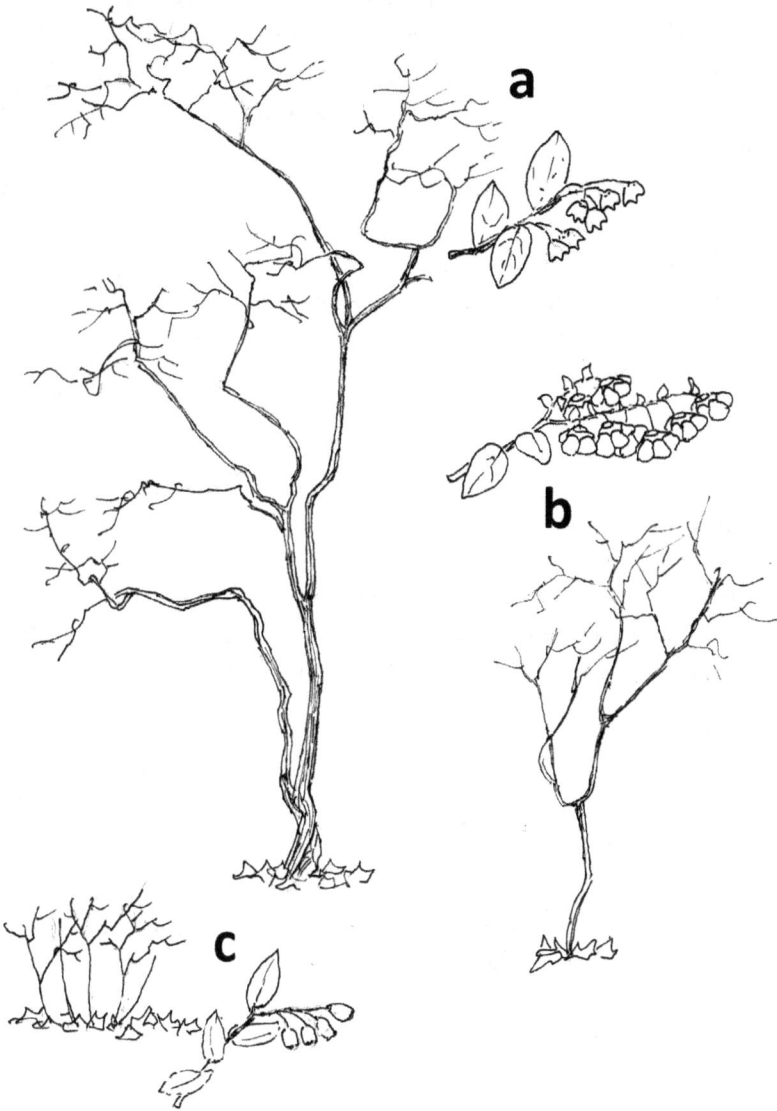

FIGURE 9.5. Winter profile and flowers of the three common species of blueberry in the Ozarks: (a) farkleberry or tree blueberry, (b) deerberry, and (c) low-bush huckleberry.

belongs to a different genus, *Gaylussacia*.) The extensive low thickets of low-bush huckleberry develop because the plant can expand by underground propagation (Figure 9.6). The underground stems are described in the technical literature as "woody lignotubers" and have features of both stems and roots. Many other trees and shrubs, such as quaking aspen in the mountains of Colorado, propagate in this way. Viewers of fall foliage on hillsides in the Rocky Mountains commonly note the mosaic of shades of gold as a way of recognizing large clusters of aspen trees that are all part of the same plant, connected by a single root system. We can do the same thing with patches of low-bush huckleberry by noting how large a section of ground has huckleberry plants coming into flower at exactly the same time or shedding their leaves in the same way in the autumn.

One other blueberry species, the deerberry, is of note in the Ozarks because its flowers are much more attractive than those of related species (Figure 9.5, b). The deerberry is a taller shrub than the low-bush huckleberry and tends to grow in isolated clumps, rather than in the nearly continuous thickets that sometimes carpet the ground in oak or pine forests. The branch tips of deerberry have a distinctive

FIGURE 9.6. Underground structure of roots and runners ("woody lignotubers") of low-bush huckleberry, which enables the shrub to propagate by underground expansion.

purple-red color, compared to the green or brown of the other native species in genus *Vaccinium*. Most species of blueberry have flowers like small pinkish-white bulbs, with a small circular opening at the outer or lower end. Deerberry has clusters of pure white flowers that flare open like delicate white bells, sometimes making a showy spring-time display. The flowers may not be as immediately conspicuous as those of more showy shrubs, but they are worth taking the time to examine up close to appreciate their delicate beauty.

Another notable member of the Ericaceae that can be found in our forests is the mountain azalea (Figure 9.7), which generally grows to a height of four to eight feet and tends to occur in fairly open forests. The mountain azalea's flowers appear in late April or early May, just as its leaves are beginning to expand. Bright pink and more than an inch across, with a somewhat funnel-shaped base and five flaring petals, these are among the most spectacular flowers produced by any shrub in the Ozarks. They are also extremely fragrant, and the pleasant odor is often sensed even before the flowers come into view. Because azaleas are adapted to thin, rocky soil and prosper where they receive some extra sunlight, you can often see their brilliant pink color aligned along the upper edges of bluffs, where they follow the line of exposed sandstone ledges. Like the dogwood, the azalea forms its flower buds at the end of the growing season, which allows hikers in fall and winter to identify places where the azalea display will be especially vibrant for viewing on springtime excursions.

Spicebush is a medium-sized or sometimes large shrub that is more often identified by its spicy scent than by its leaves or flowers (Figure 9.8, a, b). All parts of the plant are highly aromatic, which can be demonstrated easily by crushing a leaf and then holding it close to the nose. Spicebush characteristically occurs in low, moist forests, often near streams. Those going cross-country in Ozark forests often find dense thickets of spicebush a real obstacle to their passage along drainages, which would otherwise seem to be the most effective route (compared to steep slopes above, carpeted with slippery layers of leaves). This passage becomes especially difficult when the spice-bush thickets are laced with greenbrier and other vines. The fruits of spicebush are olive-shaped, bright-red drupes that are about a half an inch in length. Many different birds and small mammals eat the

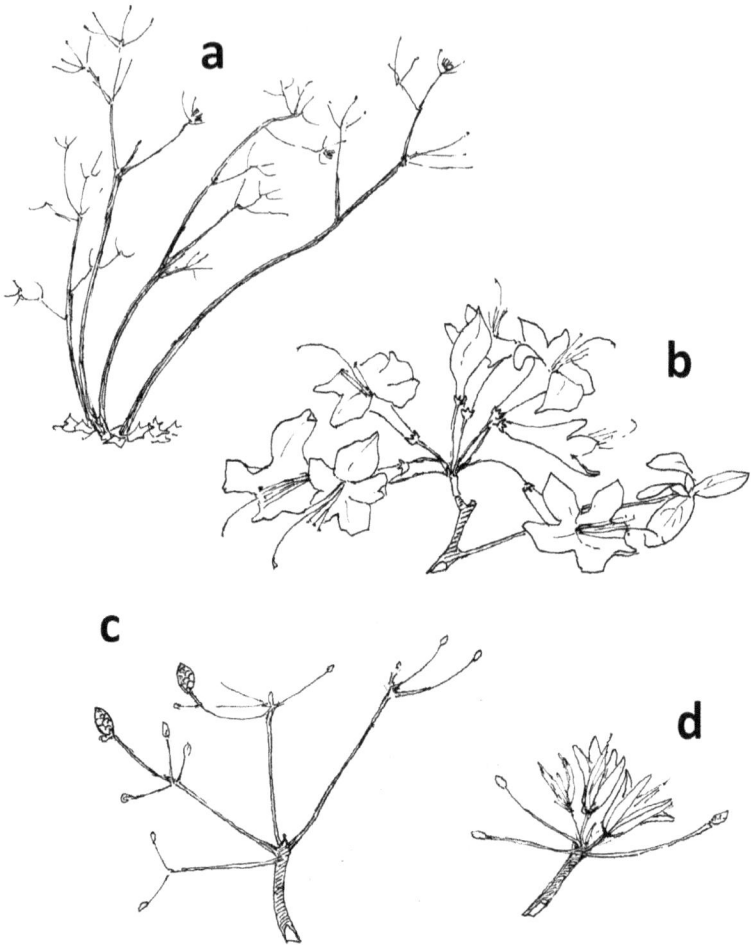

FIGURE 9.7. Mountain azalea: (**a**) profile of the multi-stemmed shrub in winter, with opened seed pods present; (**b**) flowers during early May; (**c**) detail of a twig, with flower buds present on tips during midwinter; and (**d**) dried-out azalea seed pods in the winter after flowering.

fruits. The crushed, dried leaves have been used by humans to brew a type of tea, and the dried berries have been ground up to produce a meat-seasoning spice.

Witch-hazel is another example of a plant that can vary considerably in size from place to place (Figure 9.8, c, d). It usually has the growth form of a shrub but can sometimes reach the height and size

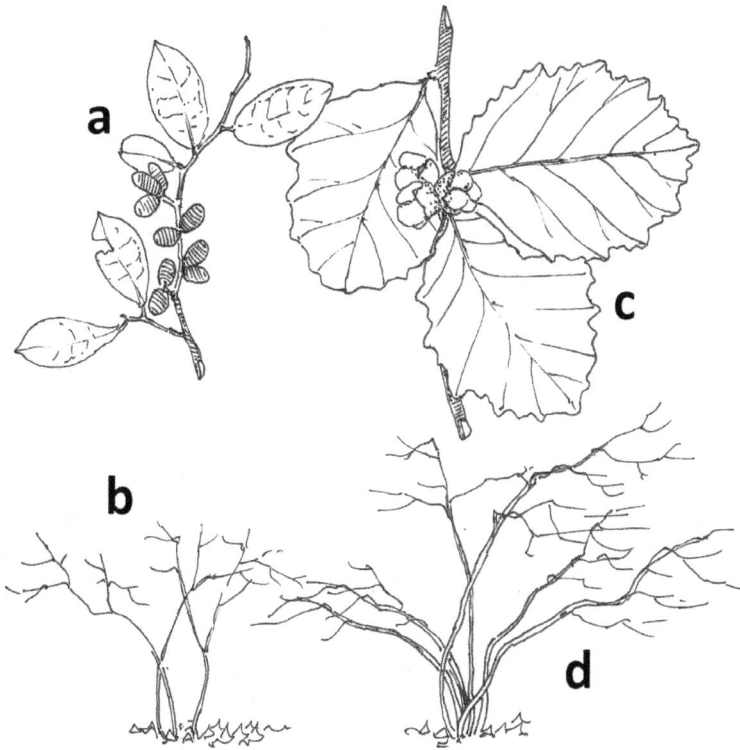

FIGURE 9.8. Growth form, leaves, and fruits of (**a, b**) spicebush and (**c, d**) witch-hazel.

of a small tree. Witch-hazel is generally associated with moist forests in the Ozarks and is readily recognized by the distinctive leaves that have a wavy margin and an asymmetrical base (i.e., one side of the base is slightly higher than the other). This plant was used for medicinal purposes by Native Americans, and extracts from witch-hazel are still used in some health-care products. The two species of witch-hazel that occur in the Ozarks are very similar except in one important respect—the time of flowering. One species (the Ozark witch-hazel) produces orange-brown flowers very early in the year, sometimes even in January. The other species (common witch-hazel) has pale yellow flowers very late in the year, often in October or November. In both species, the flowers are relatively small and not particularly colorful, which means that they are easily overlooked. The one distinctive aspect of the spring-flowering variety is its overpowering fragrance (discussed in detail in Chapter 10). The tree's common name comes from the use of its crooked branches as "witch sticks" in divining for water.

All the shrubs discussed so far occur commonly within the forest interior, but sumac tends to be found at forest edges and in relatively open areas, including thickets (Figure 9.9). Two species of sumac are widespread in the Ozark region—winged sumac and smooth sumac. Both have large leaves (sixteen to twenty-four inches long in winged and only slightly smaller in smooth) that are divided into a series of leaflets. In winged sumac, the leaf stalk to which the leaflets are attached is the "winged" part (i.e., a thin extension runs along each side of the entire leaf stalk). This "wing" structure is completely lacking on the leaf stalks in smooth sumac. Winged sumac has shiny, oval leaflets, whereas smooth sumac has longer, more pointed and finely toothed leaflets that are slightly hairy. Both species produce flowers in a compact cluster at the very top of the plant. The flowers are small, greenish-yellow, and not particularly conspicuous, but the fruits (best described as dry, hairy drupes about a quarter of an inch in diameter) are dark red when mature and thus readily noticed. The ripe fruits of smooth sumac are sometimes used to make a beverage called "Indian lemonade," and those of winged sumac have been used to produce a red dye. Both types of sumac were considered to have medicinal properties by the Native Americans.

Vines are plants that don't grow erect but instead have climbing or trailing stems. Woody vines are not particularly common in the forests of the Ozarks. The most noteworthy examples are various species of wild grape, Virginia creeper, and poison ivy (Figures 9.10 and 9.11). All of these are more likely to occur in disturbed forests than in undisturbed forests, and Virginia creeper and/or poison ivy can form an almost complete ground cover in some highly disturbed areas. Most people are aware that poison ivy has a three-part leaf and that, for most people, direct contact with any part of the plant produces an itching, irritating, and sometimes painful rash. Because the leaves of poison ivy vary in size, shape, and whether the leaf surface is dull or shiny, the well-known phrase "leaves of three, let it be" is a pretty good rule to follow. Leaves may not be evident on the larger stems of poison

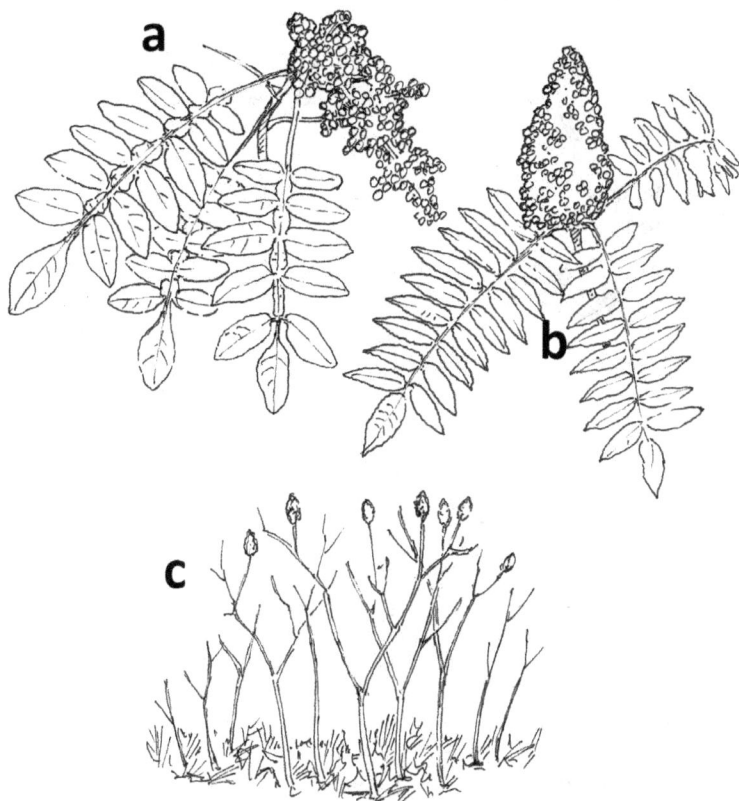

FIGURE 9.9. Typical leaves and berry clusters of (**a**) winged sumac and (**b**) smooth sumac. Both species form "domed" thickets as they spread by means of underground runners, in areas subject to disturbances such as logging or road construction. Berry clusters remain on branch tips through the winter.

ivy that grow up the trunk of a tree, but these stems can be recognized by their distinctive "hairy" appearance due to the presence of numerous aerial roots that serve to hold the stem in place.

Virginia creeper can have a similar growth form (i.e., producing a stem that often extends up the trunk of a tree) and might be confused with poison ivy if one is unaware that the stems of the former are never hairy and that its leaves have five parts instead of three (Figure 9.10, a, b). Both Virginia creeper and poison ivy produce clusters of small berries in late summer, but the former bears black berries with a

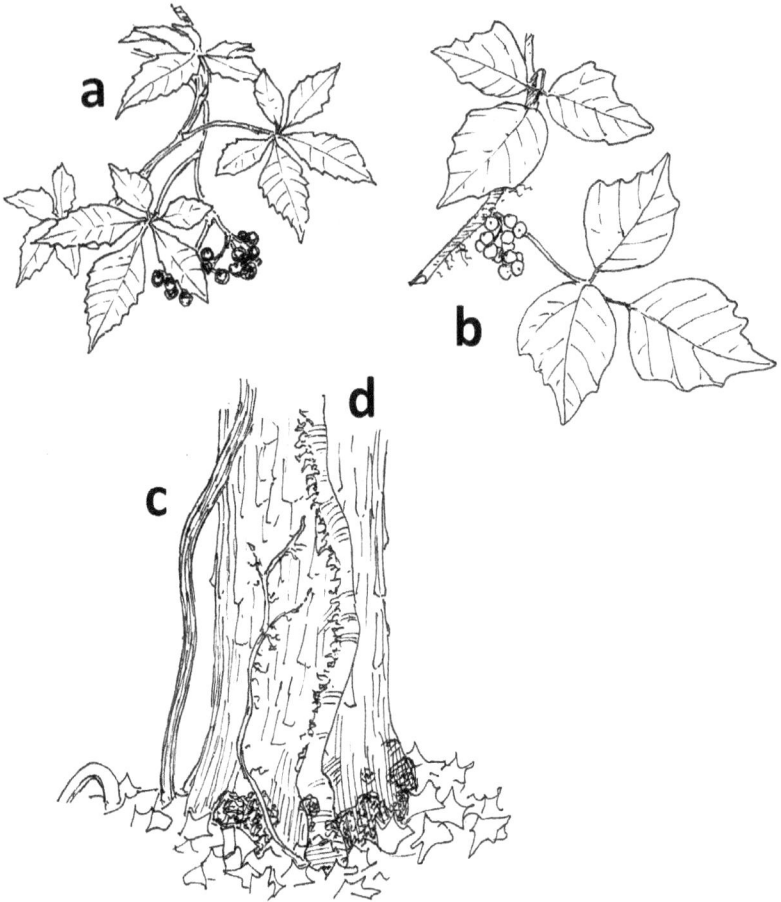

FIGURE 9.10. Leaves, stems, and fruits of (**a**) Virginia creeper and (**b**) poison ivy. Virginia creeper vines have a medium-brown striated bark (**c**). Poison ivy clings to tree bark by means of its swarms of thin, fibrous roots (**d**).

bluish blush while the latter's berries are white. The leaves of Virginia creeper turn a brilliant red in the fall, showing that it is a relative of the Eurasian "Boston" ivy seen covering the hallowed brick walls of university buildings in the eastern United States.

Grapes belong to the same family (Vitaceae) as Virginia creeper, but their leaves are not divided into leaflets (Figure 9.11). The leaves can have three or five lobes and the leaf margin is toothed. There are seven species of grape known in the Ozarks, two of which are fairly

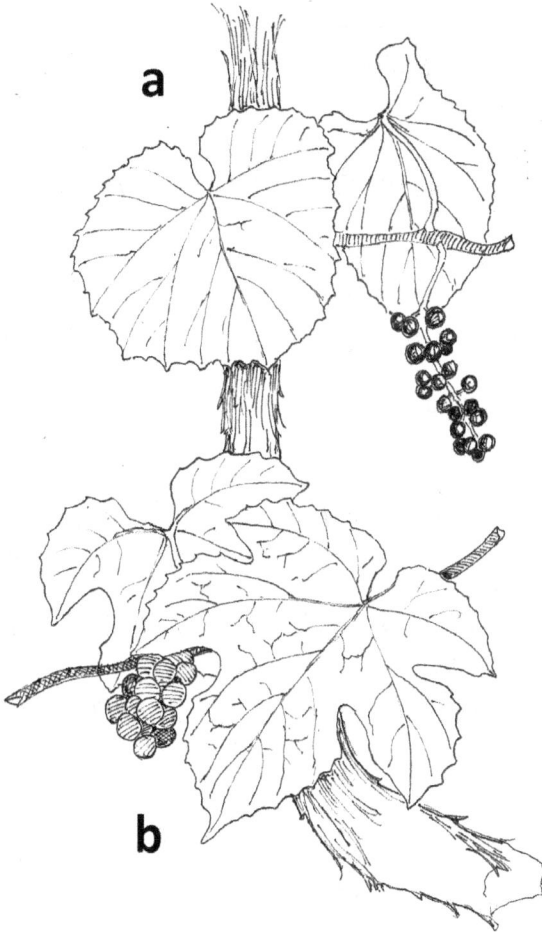

FIGURE 9.11. Leaves and fruits of (**a**) winter grape, with its small, shiny black fruit; and (**b**) summer grape, with its larger, bluish-blushed fruit. A typical mature vine with dark brown, fibrous bark is seen in the background.

common. Summer grape is characterized by rather large leaves (four to eight inches long) that can range from somewhat heart-shaped to five-lobed. The underside of each leaf is densely covered with cobwebby hairs. The fruits (grapes) are relatively large, have a bluish blush, ripen in late summer, and are sweet. Winter grape has smaller leaves (two-and-a-half to six inches long) that lack lobes, and its fruits are smaller, shiny black, and not as sweet. Both types of grape are fed upon by many different birds and other animals.

Grapevines are a familiar feature of even older, long-established forest stands. But we shouldn't think of grapevines as climbing up into the tops of large forest trees, where they are often seen (Figure 9.12). Instead, the vines and the tree probably started out at the same time, after a disturbance opened the forest, and grew up together. Indeed, grapes commonly seed into young, disturbed locations and begin to entwine themselves into small trees. During the exclusion stage of forest development (Chapter 6), a young vine can find a holdfast in one of the emerging dominant trees by growing laterally through the stand of young trees—which may require traveling some distance from its point of germination. We can see this if we look carefully: a grapevine as thick as your forearm may appear to be climbing into a large oak, but following it you find that the vine winds around for some distance on the ground before it begins its ascent into the large tree. The vine extends from where it first grew into a sapling that was shaded out to where it found its way into the crown of one of the other saplings that would eventually win the competitive race.

For those hikers who enjoy relaxing with a nice glass of wine after a vigorous walk, North American grapes should hold an especially warm place in their hearts. That's because our wild grapes saved the wine industry at a time when the famed grape-growing regions of France were faced with disaster. Starting about 1863, a virulent disease had destroyed a major portion of French vineyards and threatened to spread throughout Europe. Careful detective work by Missouri State Entomologist Charles Riley showed that the blight was caused by an accidentally introduced American aphid that injected a root-killing toxin into the vines. The wine industry as we know it was saved by providing blight-resistant roots from American grape species onto which proven wine varieties could be grafted in European vineyards.

One group of semi-woody vines, the greenbriers (Figure 9.13), warrant discussion because they are exceedingly common and tend to be noticed in a rather unpleasant manner. Greenbriers are monocots, a large and diverse taxonomic assemblage of plants that also includes the grasses, sedges, and lilies. As a group, monocots lack the capability to produce true secondary growth (i.e., essentially what we recognize as wood), so their stems cannot be considered woody (although some are exceedingly tough). Common greenbrier occurs in thickets, along forest edges, and in open (especially disturbed) forests. The leaves are broad, rounded or heart-shaped, and green on both the upper surface and the underside. The stems are green, more or less four-angled, and

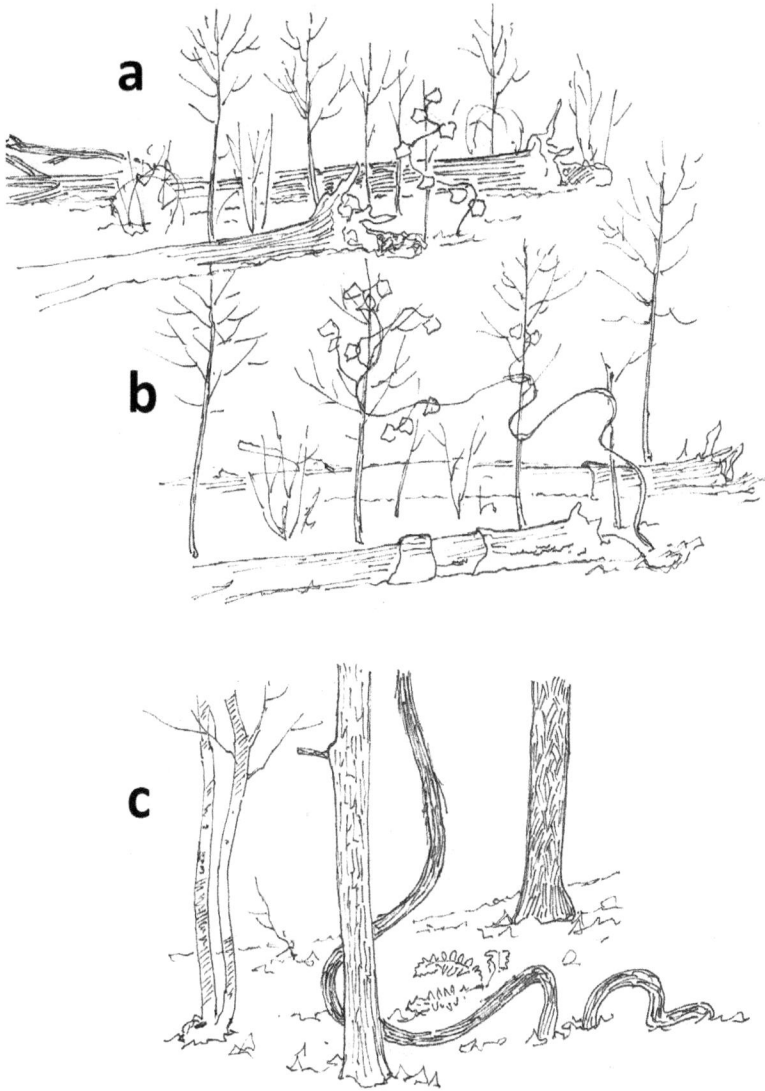

FIGURE 9.12. Development of a mature grape vine. A grape seedling begins to climb onto a tree seedling in a disturbed forest (**a**). The vine moves laterally upward to enter the crown of one of the larger saplings in the area (**b**). Eventually, the vine grows into the crown of a mature, canopy-dominant oak, but inspection shows the vine coiled for some distance on the ground from the point where it originated.

armed with prickles that can inflict damage to the unprotected skin of an arm or leg. Cat greenbrier is similar, but the lower surface of the leaf is much lighter in color than the upper surface. As already mentioned, the dastardly combination of spicebush thickets and greenbrier entanglements can really make life difficult for those who dare to walk cross-country in Ozark forests.

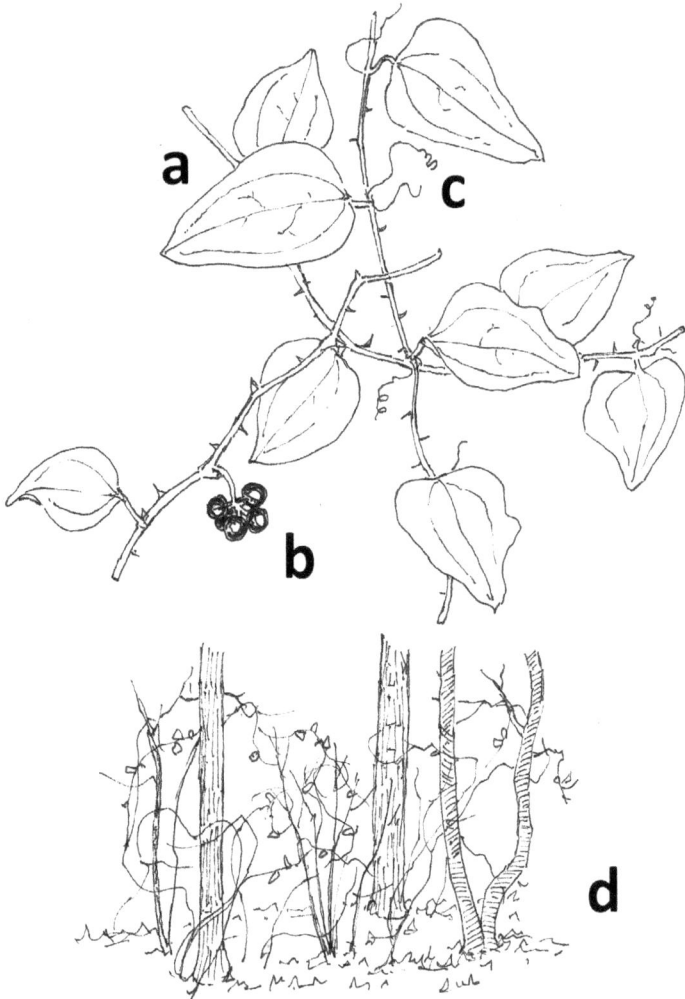

FIGURE 9.13. Leathery green leaves (**a**), fruits (**b**), and tendrils and vines (**c**) of greenbrier, and a typical, dense greenbrier thicket in a young, disturbed forest (**d**). Greenbrier is partially evergreen, with some green leaves present into late winter.

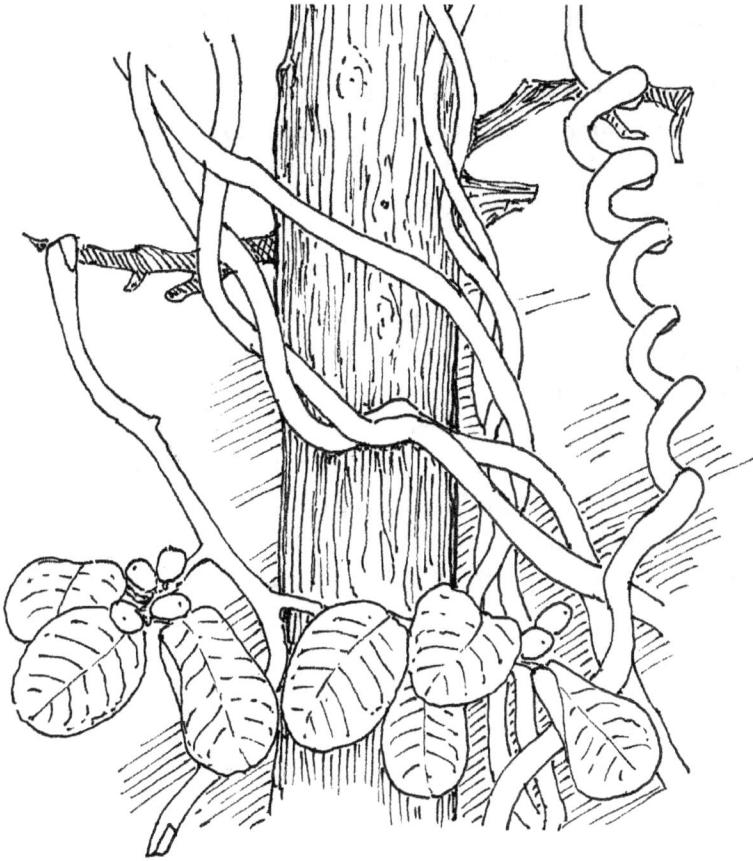

FIGURE 9.14. Rattan vine with the typical bundle of relatively thick stems, smooth green bark, oval leaves, and unripe green berries that will turn bluish-black when mature. Note the corkscrew shape of the vine stem, which shows that it was once entwined with a now dead tree branch.

Another vine, less common but still widespread, is sometimes misidentified as an overachieving version of the pernicious greenbrier. This particular vine has a glossy green or green-brown stem that can be as much as a full inch in diameter and sometimes reaches the very tops of trees. The stems of these vines look very much like those of greenbrier, but they are clearly much thicker than any greenbrier vine you will ever see, and they lack the prickly thorns. The plant in question is rattan vine (or "supple-jack"; Figure 9.14), a member of the buckthorn family. All other North American species in this family grow as shrubs or small trees, a relationship that helps confound the amateur botanist, because rattan vine tends to be listed with buckthorn shrubs and not in the section of an identification book that includes the other common species of vines. In addition to being mistaken for a giant greenbrier, rattan vine is noteworthy in two ways. First, the tough but supple stems of the vine were ideal for the wickerwork of pioneer crafters who settled in the frontier Ozarks. More relevant for the hiker today is the unusual growth form this vine can achieve. Unlike grape and Virginia creeper vines, rattan vine tends to occur in dense bundles of cable-like stems. In the process of growing into these bundles, young vines tend to support themselves by twisting around other vines or the stems of small trees. As the stem of the vine grows larger over time, the stem around which it was entwined may have decayed away. This results in a striking corkscrew form that can really catch the hiker's attention, causing the forest scene to resemble the kind of exuberant growth typical in photographs of the Amazonian rainforest.

CHAPTER 10

Species of Special Interest

Many trees and other plants found in Ozark forests are of special interest because of their distinctive appearance, the body of folklore associated with them, or the simple fact that they are rarely encountered. In some instances, the plant in question is much less common today than it was in the past. One noteworthy example is the Ozark chinquapin (a variety of chestnut), which stands out in Ozark folklore and probably was once a major source of food for wildlife (Figure 10.1). The chinquapin (which should be carefully distinguished from chinquapin oak, a common large forest tree in the Ozarks today) was eliminated as a nut source when the introduced chestnut blight reached our area after 1950 (see Chapter 13). The tree still survives by sprouting from its roots, and older Ozark residents cherish memories of harvesting chinquapin nuts in the fall. Some may question the veracity of those memories of abundant nut harvests from a relatively uncommon forest tree. But recent studies by an Ohio University doctoral student, based on limited data samples available in blight-free outliers of American chestnut, show that the latter species' seed crop could have been at least a hundred times as great as the typical acorn crop from red oaks. The Ozark chinquapin, a relative of the American chestnut, may have produced comparable amounts. The chinquapin's nut crop would also have been a reliable yearly food source, because the tree blooms in late May, when all chance of frost damage to the flowers and young fruits is long past. Such regular nut crops might seem a puzzle when compared to the rodent-suppression strategy of irregular nut crops in oaks (see Chapter 3). We speculate that chestnut and chinquapin seedlings are so shade tolerant that they can survive indefinitely on the forest floor. For that reason, these nut trees want their seed to be spread as far and wide as possible, with the assumption that the few nuts that escape consumption will be enough to ensure regeneration.

FIGURE 10.1. Ozark chinquapin foliage and male catkins (left) with the logo (right) of the Ozark Chinquapin Foundation, a nonprofit organization dedicated to the restoration of chinquapin to the Ozarks. The logo shows the nuts and burs of the tree and the wildlife that used to feed on its nuts. Male catkins appear when the tree is still relatively small and thus are a common sight. Female flowers that can develop into nuts are found on older trees that have escaped blight for a decade or more and thus are very rare.

We can infer a lot about our lost chinquapin tree from studying the available data. Although it was once considered a variety of the shrubby Allegheny chinquapin, DNA studies show that Ozark chinquapin is an ancient lineage that is more distantly related to American chestnut and Allegheny chinquapin than those two are related to each other. We can also see that our chinquapin was a large forest tree from the easily recognized trunks of individuals killed by the original appearance of blight. In fact, the stature of the tree can be reconstructed by carefully measuring the dimensions of long-dead mature trees and then "fleshing out" the fine-scale branches in the tree's crown using the shape of the largest surviving chinquapin sprouts found today (Figure 10.2). These results show that Ozark chinquapin was fully as large as surrounding trees on the relatively dry and exposed ridgetop habitats that the tree preferred, and it was growing as a single, upright stem with a rather crooked growth form and numerous small stems arrayed around the base of the main trunk. This is certainly far different from the bush-like stature associated with Allegheny chinquapin.

FIGURE 10.2. Reconstruction of the size and growth form of old-growth Ozark chinquapins, based on the measurement of logs representing trees killed by the original pandemic of chestnut blight. The crowns have been filled out using the branching pattern of the largest chinquapins that can be found to have temporarily escaped blight in today's forest.

"Witness tree" data from early land surveys (Chapter 5; Figure 5.2) show that chinquapin was about as common as ash and maple in the early nineteenth century. Studies of the remains of old, blight-killed chinquapin show that these trees once grew in groves of a few to about a dozen trees on dry ridgetops. Interestingly, very few of these original trees managed to resprout from their base after the initial arrival of blight. The many living chinquapin sprouts found in the vicinity of the remains of these trees were seedlings established before blight cut off the seed source. These "old seedlings" continue to resprout after rounds of stem girdling by blight and can probably survive indefinitely—until the researchers who are actively attempting to breed a blight-resistant variety produce trees that can be reintroduced to reestablish Ozark chinquapin as a viable forest tree.

Another plant that is undoubtedly less common today than it was a century ago is ginseng, or more properly American ginseng (Figure 10.3), a perennial herbaceous plant native to broadleaf forests of the eastern and central United States. A similar plant occurs in eastern Asia, and the same is true for a number of other plants, including such common examples as dogwood, mayapple, Virginia creeper, sassafras, and tuliptree, all of which are represented by very similar species in both regions of the world. This pattern of distribution, first described in detail by Asa Gray, a botanist from Harvard University (1810–1888), has been referred to as the "Asa Gray disjunction." The similarity between some American forests and those of China is a modern echo of a single, Northern Hemisphere–wide deciduous forest (known as the Arcto-Tertiary Flora) that stretched all the way across the top of the globe some sixty million years ago, when the northern continents were joined and the climate was warmer. Although cooler and drier climates in Scandinavia and Siberia have since eliminated these biologically rich forests from the lands between America and China, echos of that ancient environment survive in these two former endpoints of the great prehistoric forest.

In China, wild ginseng has been harvested for thousands of years for its purported curative properties. According to an ancient Chinese legend, early emperors proclaimed it a panacea to be ingested or used in lotions and soaps. The name of the genus (*Panax*) is based on the Greek word *panakeia* (literally "universal remedy"). The common name "ginseng" is derived from the Chinese term *jen-shen*, which means "in the image of a man." Indeed, the root can look almost as if it has arms and legs (Figure 10.3). While ginseng roots of any shape are highly prized, those that resemble the human body are considered especially desirable and, as a result, have a considerable market value.

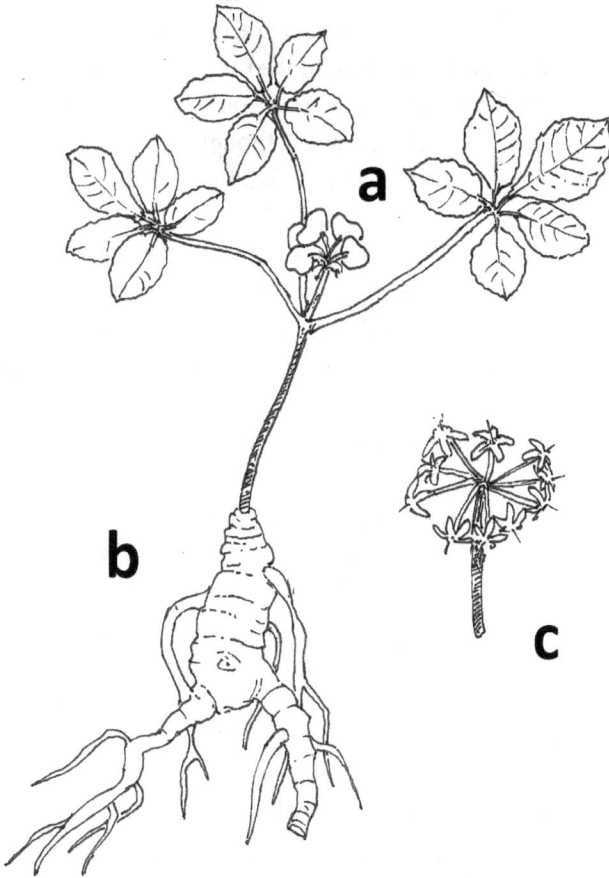

FIGURE 10.3. American ginseng: (**a**) foliage and berries, (**b**) a root that looks almost human because of limb-like projections from the trunk, and (**c**) flowers.

Because the plants are so very similar, American ginseng is collected here and marketed in China. Indeed, dried ginseng roots were among the first marketable plant products to be shipped from the United States to China, as early as the latter part of the eighteenth century. Today, about eighty thousand tons of ginseng are sold worldwide. Unfortunately, over-collecting has decimated ginseng populations in many portions of its former range in the central and eastern United States, but it is still possible to find the plant in places in the Ozarks characterized by rich, moist soils. Ginseng is easily recognized from its simple erect stem with three leaves, each divided into five finely toothed leaflets, and (in late fall) its rounded cluster of red berries.

Orchids have always held a fascination for plant enthusiasts and gardeners that is compounded by their rarity in our landscape. Orchids were never especially common in the Ozarks, but their numbers have certainly been reduced as a result of the loss of suitable habitat. The orchid family (Orchidaceae) is thought to be the largest single family of flowering plants, but one defining characteristic of the family is that particular species of orchids tend to be uncommon or rare. The reason for this is that orchids are the most mycorrhizal of all plants: they are so dependent on their fungal partners that the chances of a new plant being established are very low (see Chapter 12). Orchids counter this by producing, on average, the most seeds of any flowering plants, but those high numbers of seeds are possible only if the energy and nutrient resources the plant diverts to the production of each seed are relatively limited. As such, when the seeds are released, they are underdeveloped in comparison to the seeds of other flowering plants. Once dispersed, a seed must come into contact with a suitable host fungus very quickly or it cannot germinate and give rise to a new plant. Presumably, this doesn't happen very often, and the result is a low population of the orchid in question. It is not unusual, when hiking through a forest, to encounter a certain type of orchid once but never see another during the entire hike.

The vast majority of orchids are tropical, and the group is not especially well represented in the Ozarks. However, about two dozen species have been recorded from the region, almost all of which are uncommon. The most abundant and widespread orchids in the Ozarks are two species of ladies' tresses and the downy rattlesnake plantain. The flowers produced by these three species are not especially noteworthy, but that is not the case for the spectacular ladyslipper orchids (Figure 10.4). The petals of their flowers are fused together to form a slipper-shaped structure (hence the common name) with an expanded pouch-like portion. This entire structure is a specialized trap for the insects that pollinate the flower. Four species of ladyslipper orchids occur in the Ozarks. These are the Kentucky ladyslipper, the small yellow ladyslipper, the large yellow ladyslipper, and the showy ladyslipper. The pouch is pink in the latter species and some shade of yellow in the other three. The showy ladyslipper is by far the rarest and most seldom encountered of these species, and there are fewer than half a dozen populations of which botanists are aware. Populations once known to exist at a number of localities are now gone because the plants were removed from the wild by collectors who wanted to transplant the showy ladyslipper to their own flower garden. The unfortunate and continued poaching of ladyslipper orchids extends to the other species as well. The people who do this may be unaware that it is almost impossible to transplant orchids because of the unique association they have with mycorrhizal fungi.

The ladies' tresses are probably the most widespread and abundant orchids found in the Ozarks, but they tend to be overlooked, largely because their flowers are white and relatively small. These flowers are arranged in a distinct spiral along the upper portion of a green stem that may appear leafless, since the basal leaves of the plant tend to have

FIGURE 10.4. The large, attractive yellow ladyslipper orchid is rare, found in only a few secluded places in the Ozarks.

withered away by the time of flowering, which takes place in late summer. Ladies' tresses occur in a variety of habitats, including along moist roadsides—seemingly a most unusual place to find native orchids. However, if one actually looks for these plants, they are not especially difficult to notice. One of the most commonly encountered species, the southern slender ladies' tresses, can reach a height of two feet.

Downy rattlesnake plantain can be found in scattered localities throughout the Ozarks, where it is typically associated with somewhat acidic soils (Figure 10.5). The third part of the common name is derived from the fact that the leaf arrangement (all of the leaves radiate outward from the stem near the surface of the ground, a condition referred to as a "basal rosette") and shape of the leaves are similar to the common plantain that probably occurs in every lawn in the central United States. However, the leaves of downy rattlesnake plantain are characterized by a distinctive network of white veins and a pronounced white stripe down the midrib, which is quite different from the uniform green of plantain. With a little imagination, the overall pattern of the leaf resembles the skin of a snake, hence the "rattlesnake" part of the common name. In late summer, downy rattlesnake plantain produces an erect flower stem bearing a series of relatively unimpressive white flowers, but we take that in stride because the leaves themselves are so intricately attractive.

During early spring, before many trees have produced their leaves, the large (often more than an inch across), deep reddish-brown flowers of pawpaw are the most conspicuous of the early-flowering woody plants (Figure 10.6). Although striking in appearance, the flower has a somewhat foul odor, which is important for attracting some of its insect pollinators, which are attracted to carrion and not to dainty flowers. Pawpaw is a small tree, commonly found on well-drained but rich soils of low areas throughout the Ozarks, usually reaching a height of no more than forty feet. However, since pawpaw is capable of forming clones, thickets consisting of multiple, much smaller stems are often encountered. Such thickets can interfere with the establishment and growth of other woody plants. A good way to see for yourself how many pawpaw stems make up a single plant is to observe the tree when it flowers in early April. The flowers can vary between individual clones, and thus you can determine how many stems in an

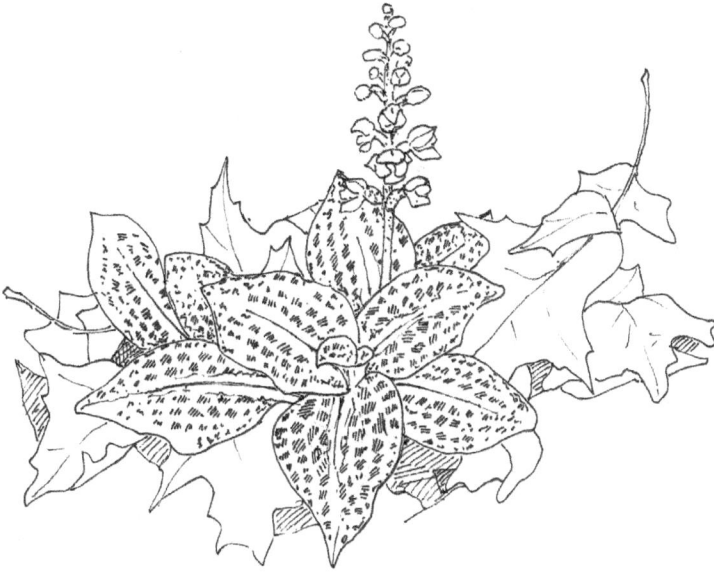

FIGURE 10.5. Downy rattlesnake plantain, showing the distinctively mottled leaf rosettes and rather plain-looking spike of small white flowers of this relatively common and widespread Ozark orchid. The basal rosette is formed where all the leaves radiate outward from the stem near the surface of the ground.

area have exactly the same shade of color and open at exactly the same time—just as you can do to estimate the size of huckleberry clones (as described in Chapter 9). The leaves of the pawpaw are simple and relatively large (up to a foot in length), and the fruit is the largest edible fruit (as much as several inches long) produced by any plant native to the central and eastern United States. The ripe fruit is consumed by wildlife and sometimes collected for human consumption. If you would like to taste one for yourself, you are going to have to get up early to beat the raccoons.

The leaves of the umbrella magnolia are even larger than those of the pawpaw, and exceptional examples can reach a length of almost two feet (Figure 10.7). Umbrella magnolia is a small tree and rarely exceeds fifteen feet in height, so, in a comparative sense, these are very large leaves for what is a relatively small tree. The leaves are clustered at the end of the stem, and the arrangement of leaves in a cluster somewhat resembles an umbrella (hence the common name).

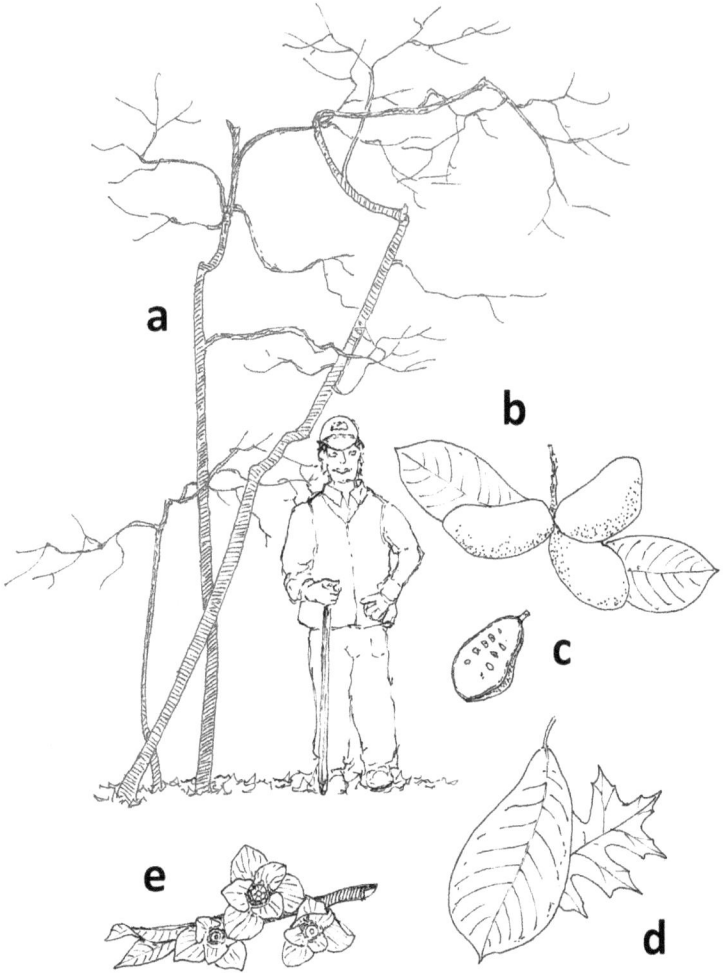

FIGURE 10.6. Pawpaw: (**a**) growth form of mature stems in a typical shaded environment (with hiker for scale; all three stems are probably part of the same plant); (**b**) twig with ripe fruit; (**c**) cross section of fruit showing seeds embedded in pulp; (**d**) typical leaf (with typical oak leaf for scale); and (**e**) twig with flowers that typically appear in the last days of March or in early April.

Umbrella magnolia is a characteristic small tree in the Appalachian Mountains of eastern North America, but its range extends westward all the way to the Ozarks. Thus, it is one species of tree that has been able to jump the Appalachian "gap" imposed by the Mississippi River (see Chapter 2), unlike other species such as tulip poplar, which never made that leap beyond Crowley's Ridge (even though they seem to do very well when introduced into our forests). The most remarkable feature of umbrella magnolia is the flower, which is huge. With their large white petals and a total diameter approaching ten inches, these flowers can be observed from a considerable distance.

Along streams and in low, moist areas of forests throughout the Ozarks, one frequently encounters a tall shrub or small tree with crooked gray branches and, when they are present, oval leaves with a somewhat scalloped margin. This description applies to two closely related species in the genus *Hamamelis*, commonly known as witch-hazel (see Chapter 9). The two species differ in two important ways: vernal witch-hazel (sometimes called "Ozark witch-hazel") is endemic to our area and flowers very early in the year (January–March), whereas the common witch-hazel is much more widespread and flowers late in the year (usually September–December). Thus, witch-hazel is both the first tree and the last tree to flower each year, although two species are involved. The flowers are easily recognized by their four slender, strap-shaped petals, which are less than an inch long and pale to dark yellow or orange (Figure 10.8). Although the two

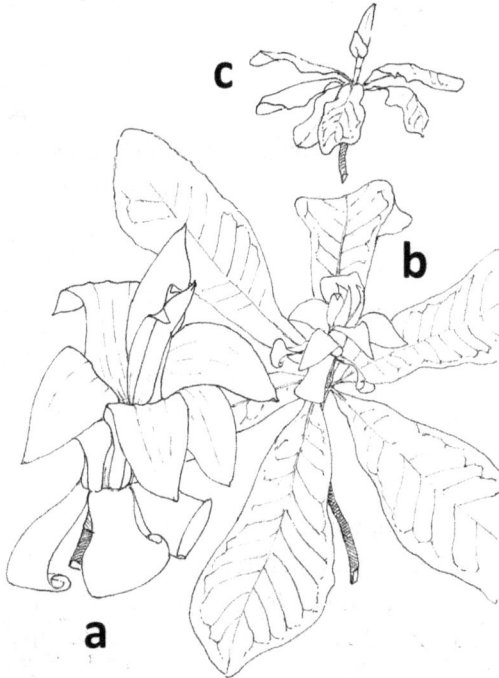

FIGURE 10.7. Umbrella magnolia: (**a**) flower, (**b**) flower embedded in the umbrella-like rosette of oversized leaves that gives the tree its name, and (**c**) flower bud before it has opened.

species' flowers are similar in appearance, those of vernal witch-hazel are highly fragrant while those of common witch-hazel are not. This remarkable fragrance makes spring-flowering witch-hazel of special interest to any outdoor enthusiast in the Ozarks. If you have not personally experienced it, be sure to seek out the plant when it flowers in February or early March. An especially good place to have this experience is on the approximately one-mile walk from the Arkansas Natural Heritage Commission parking lot to the Kings River Falls viewing area. The riverbank is lined with an almost continuous strip of witch-hazel, and the fragrance when the shrubs are in full bloom can be almost overwhelming. Because of this trait, the Ozark endemic has recently gained attention from the horticultural world and is now available through specialty nurseries.

Fringe tree is another type of small tree sometimes found in the same habitats as witch-hazel (Figure 10.9). For most of the year, a fringe tree is not particularly conspicuous, although it is fairly easy to recognize from the leaves, which are simple, opposite, ovate to oblong, and relatively large (as much as seven inches long and often more than three inches wide). Few other small trees found in the Ozarks have opposite, simple leaves this large. Fringe tree becomes much more conspicuous when it is in flower. In fact, it is hard to imagine any tree that is more spectacular in appearance when flowering occurs. The flowers are pure white, slightly fragrant, and occur in showy masses that cover the entire tree, and it is impossible not to notice the tree as it stands in stark contrast to its surroundings. Fringe tree grows wild over much of the central and eastern United States but also has been planted as an ornamental in some urban settings, including parks. You can see fringe trees in bloom in the Ozarks where the Buffalo River Trail winds along the ledges above the Buffalo National River corridor.

By definition, a shrub is a small woody plant, but there are some plants widely recognized as shrubs that can be rather tall. This is certainly true of farkleberry (aka "sparkleberry" or "tree huckleberry"), which is capable of reaching a height of thirty feet. Farkleberry is a member of the genus *Vaccinium*, which includes the blueberries and cranberries (Chapter 9; Figure 9.5), neither of which typically grow very tall (most reach a height of two feet or less). Farkleberry can thus be considered a giant member of the group. However, like

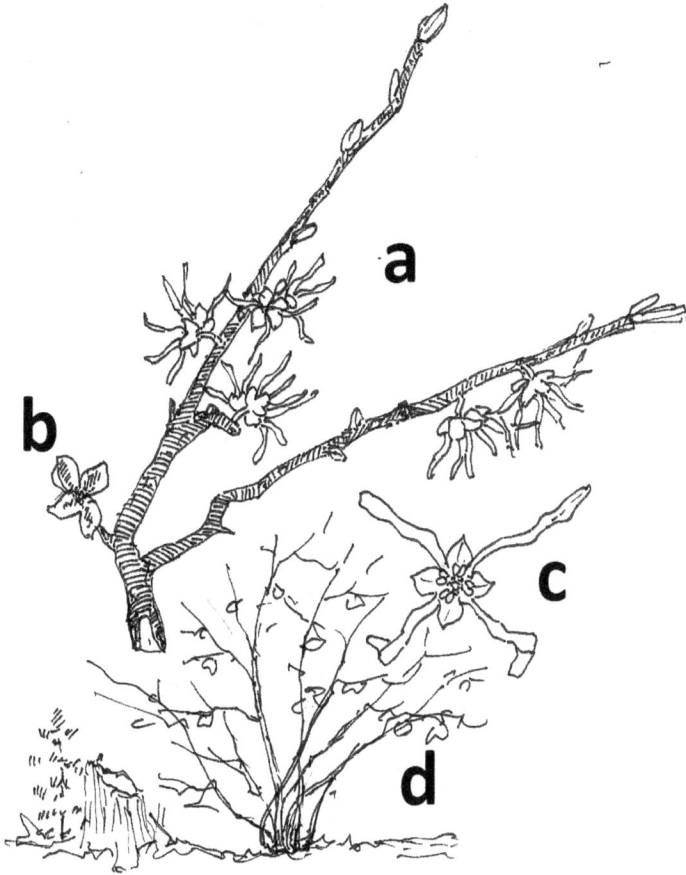

FIGURE 10.8. Vernal witch-hazel: (**a**) flowering twig; (**b**) opened seed capsule from previous season; (**c**) detail of flower structure; and (**d**) profile of the shrub (with stump and pine sapling for scale), which often retains some shriveled brown leaves through the winter.

other *Vaccinium* species and most other plants in the same family (Ericaceae), it is confined to soils that are fairly acidic, which limits its distribution in the Ozarks but makes it an important component of glades developed on acidic rock types such as sandstone and chert. It is native to rocky uplands in Arkansas, Oklahoma, and Missouri, with a range much more restricted and local than that of other blueberry species. The farkleberry often has an elaborately contorted growth

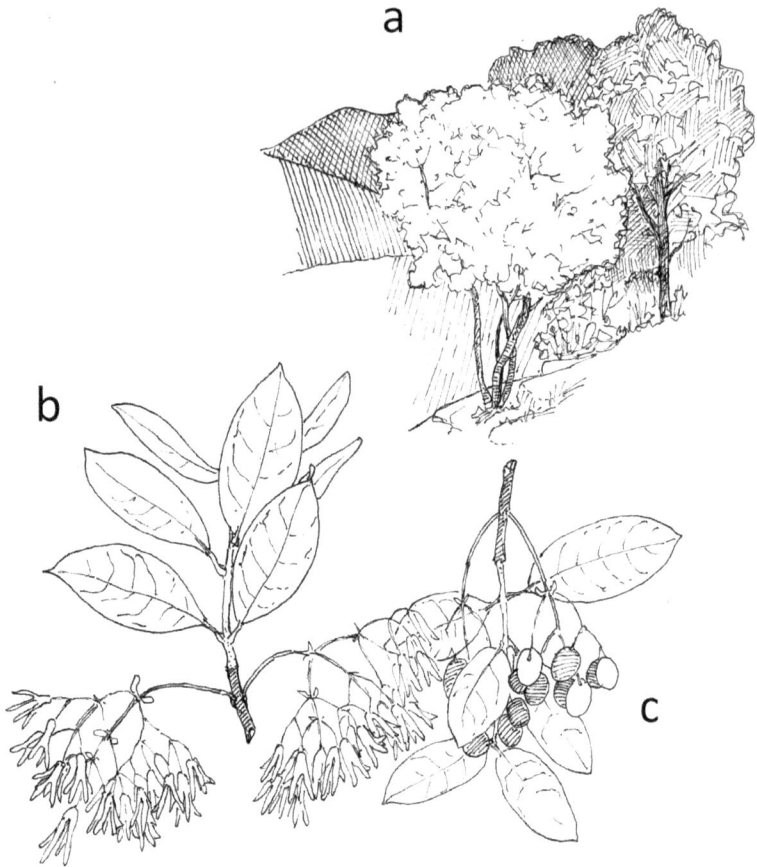

FIGURE 10.9. Even though the fringe tree blooms in late spring (**a**), when its leaves have already emerged, the large and dense clusters of four-petaled flowers make a beautiful white cloud as impressive as any dogwood; detailed are a flowering twig (**b**) and the blue-black fruits produced in early autumn (**c**).

form that makes it an interesting and important part of Ozark scenery (Figure 10.10). Hikers can be pleasantly surprised when they take the time to notice the way in which the orange-brown trunks of the large but shrubby farkleberry clumps twist back on themselves over and over, in contortions one would not have thought possible for a simple plant.

American holly is a familiar shrub to most people, if only because its spiny, evergreen leaves and bright red berries are often used as

FIGURE 10.10. Example of a scenically contorted farkleberry shrub growing on a sandstone ledge in the Ozark National Forest.

decorations in the Christmas season. Although widely planted as an ornamental, the natural range of American holly is restricted largely to more southern portions of Arkansas, but it is sometimes possible to encounter the plant in the Ozarks. However, there is another type of holly that is native to the Ozarks, much more widespread, and deciduous instead of evergreen. This is the possum haw (aka "deciduous holly"; Figure 10.11, a), a large shrub or small tree that occurs in generally moist habitats, such as along streams, throughout the entire

Ozark region. The leaves of possum haw are not spiny, but instead have a wavy-toothed margin. When its characteristic clusters of red fruits are not present, possum haw is a relatively nondescript plant and easily overlooked in the forest. Hikers are sometimes surprised to see the dramatic blaze of red that a possum haw, enveloped in a sparkling cloud of bright red berries, can provide in the drab winter landscape. The same shrub may have gone unnoticed on previous winter outings when it had been unable to produce a large berry crop because of insufficient pollination or some other interference.

Possum haw isn't the only large shrub or small tree with "haw" as part of its common name. An old English word for "hedge" (i.e., closely growing bushes or shrubs), "haw" is not specific to any particular taxonomic group of plants. Indeed, black haw (Figure 10.11, b) belongs to an entirely different plant family than possum haw, although the growth forms of the two "haws" are fairly similar. The leaves of black haw superficially resemble those of cherry, and the specific epithet (*prunifolium*) refers to this fact (cherries belong to the genus *Prunus*). However, they have an opposite arrangement on the stem, whereas the arrangement of leaves in cherries is alternate. Black haw produces blue-black, berry-like fruits that, like cherries, have "pits." Black haw berries often have such a pronounced powdery blue that they appear to have a strikingly bright blue color, and a small tree sagging under the weight of its fruit crop can be quite an interesting sight. A source of food for wildlife, the fruits are sometimes collected by humans for use in jellies and jams. The bark of black haw also has a distinctive, pronounced checkered pattern that many hikers find especially interesting.

Black locust is a common tree throughout the Ozarks (Chapter 4; Figure 4.9). Each of its compound leaves consists of seven to nineteen thin, rounded leaflets. Both the relatively high number of leaflets and their shape distinguish this species from other trees with compound leaves. And while walnut, hickory, and ash can be found growing anywhere in a forest, black locust tends to occur only at forest edges or in open areas. This is because it is exceedingly shade intolerant, which means that younger individuals (for example, seedlings and saplings) cannot survive in low light conditions. Black locust belongs to the family Fabaceae, which also includes such familiar garden plants as peas and beans. The members of this family have two outstanding

FIGURE 10.11. Two attractive and noteworthy Ozark shrubs: Possum haw has deciduous leaves and bright red berries in midwinter (**a**). Black haw has clusters of plum-like, blue-black berries in late summer (**b**) and distinctive, yellow-brown, pebble-grained bark (**c**).

characteristics. First, the fruit they produce is a pod (or legume; hence, these species themselves are often referred to as "legumes"), and black locust has flattened pods three to five inches in length that are often conspicuously present on mature trees. The second characteristic, though much less apparent, is ecologically very important. The roots of legumes establish a mutualist relationship (i.e., both partners benefit) with certain types of nitrogen-fixing bacteria in the soil. These bacteria are capable of taking nitrogen gas from the atmosphere and converting it to ammonia. The nitrogen gas itself cannot be utilized directly by living organisms, but when it is combined with the element hydrogen to become ammonia, this altered (or "fixed") form can be incorporated into the various biological molecules necessary for life. For example, nitrogen is an essential component of the structural proteins that make up living organisms. There is little question that black locust and various other legumes play an important role in the terrestrial ecosystems in which they occur.

Black walnut (see Figure 3.3) is a common tree throughout both the Ozarks and the rest of Arkansas, but the closely related butternut (or "white walnut") is only rarely encountered (Figure 10.12). Butternut tends to be a smaller tree than black walnut, and the nut is oblong in shape (not round like that of black walnut) and irregularly ribbed once the husk has been removed. Anyone who has handled black walnut husks knows how readily the sap can stain hands black. Butternut produces a similar stain with a tan or light-brown color. Folklore has it that Confederate volunteers stained their own homemade uniforms using dye made from butternut husks. More likely, these volunteers were just identified with that tree from the already tan color of their issued uniforms. Butternut was once relatively common throughout the eastern and central United States but has undergone a major decline over the past century. Although much less publicized than that of the American chestnut, this decline is similar in that the ultimate cause is a serious disease caused by a fungus—in this case, the "butternut canker." The fungus is easily transmitted from tree to tree, and diseased trees usually die within several years. In some areas, at least 90 percent of the butternut trees have been killed. Interestingly, black walnut appears to be completely resistant. Unfortunately, it appears that little can be done to prevent the almost complete demise of the butternut.

Today, eastern red cedar (often simply called "red cedar") is common and sometimes abundant in certain areas of the Ozarks, but this was not always the case. Note that the name "cedar" can lead to botanical confusion, because this cedar is a species of juniper and not at

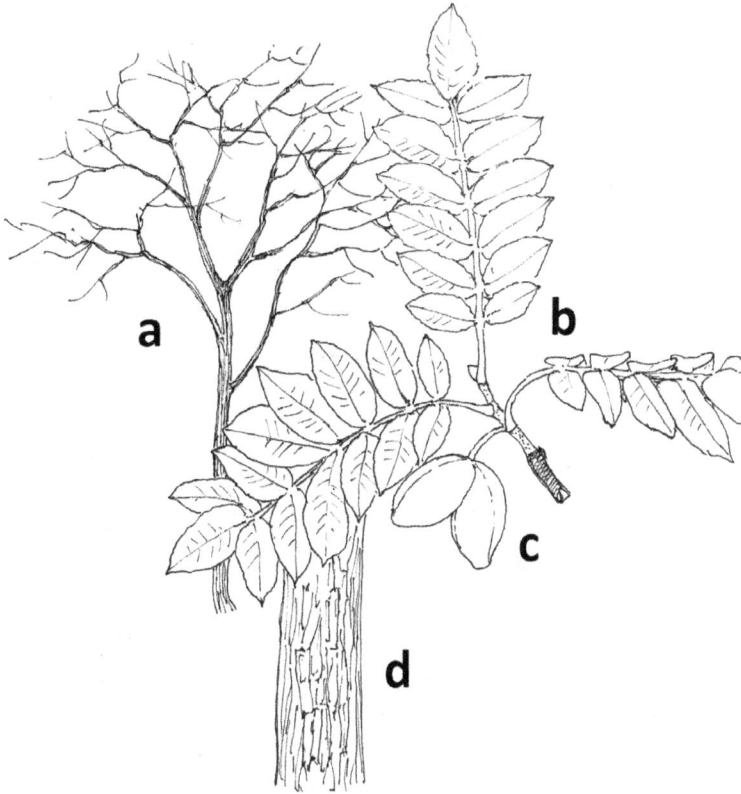

FIGURE 10.12. Butternut in profile, with typical (**a**) spreading crown, (**b**) leaves, (**c**) nuts, and (**d**) bark pattern.

all related to true cedars, which belong to an entirely different genus of conifers. Prior to human settlement of the region, red cedar was likely confined mostly to special and rather limited habitats such as cedar glades and rocky bluffs (see Figure 2.17C). Even as late as the 1940s, aerial photographs taken in portions of northwest Arkansas show little evidence of red cedar. However, the species has expanded greatly since then, primarily as a result of fire suppression (red cedar is highly susceptible to fire) and changes in abandoned fields once used for agriculture. These two factors have allowed red cedar to expand its range well beyond the habitats in which it once occurred. One important factor is that old fields are often subject to periodic grazing in the late stages before they are entirely abandoned, and the prickly foliage of young red cedars is much less palatable than the other vegetation and tree seedlings that may have begun to seed into the pasture. Once livestock are excluded from a pasture for good, our burgeoning deer population takes over and the effects are the same. Left unmanaged, cedar has the potential to form a dense and almost monodominant community at a particular locality, which has a considerable negative impact on overall biodiversity. Few other plants can survive beneath the canopy cover of a red cedar forest, and this type of ecological setting is not attractive to most species of wildlife.

Individual red cedars that occur on exposed rocky bluffs, such as those found along the Buffalo River, can reach a considerable age and develop scenic and even artistic shapes (as illustrated by the specimens in the foreground of Figure 8.6). Long-lived cedars evolve from an initially columnar seedling into a mature tree that then suffers crown decline and eventually presents a contorted appearance, with stark skeletons of dead branches and a single strip of viable bark cambium winding up the otherwise dead trunk (Figure 10.13). The maximum age ever recorded for red cedar is more than nine hundred years, and individual trees exceeding two hundred years are not uncommon. This is true only for trees found in special ecological situations; red cedars in the monodominant communities described above are usually no more than about sixty years old. Tree cores extracted from old-growth individuals of red cedar can be analyzed to provide a record of the environmental conditions (including rainfall) to which the tree was subjected over a period of several centuries. Some of these trees were alive well before the effects of European settlement became apparent in the Ozarks. Climate reconstructions based on data from red cedar tree-ring studies by the University of Arkansas Tree-Ring Laboratory hint that Arkansas has experienced droughts in the past millennium that were much more severe than even the worst "dust bowl" years of the twentieth century.

The term "living fossil" is a most appropriate term to apply to one plant found in the Ozarks, the shining firmoss (or "shining clubmoss"; Figure 10.14). This species is rather rare in the Ozarks, though often common in high-elevation coniferous forests of the Appalachians. The group of plants to which shining firmoss belongs (the lycopods) has

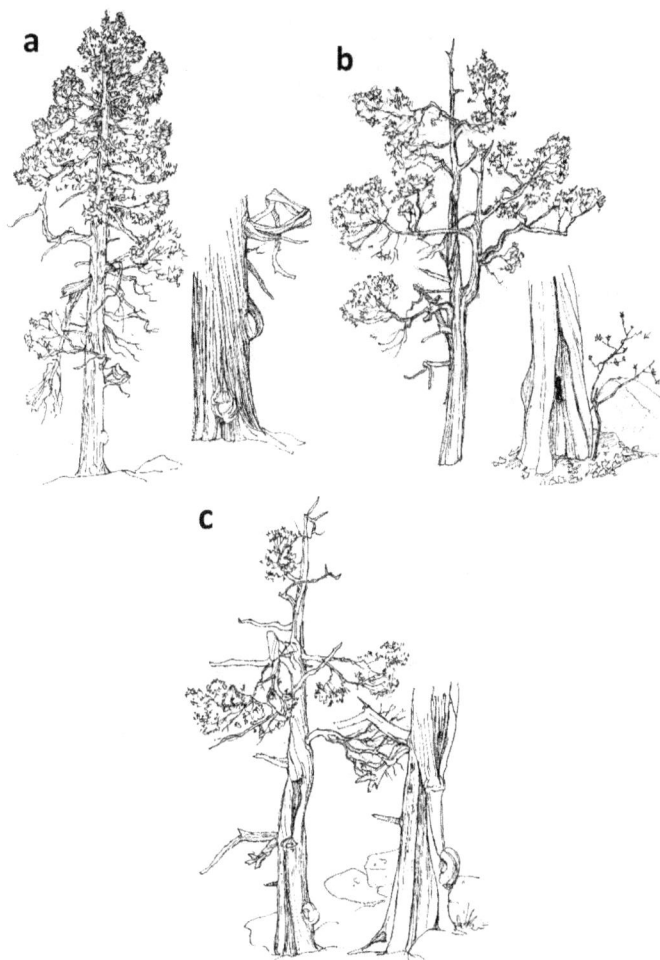

FIGURE 10.13. Life history of a red cedar, from its prime (**a**) to signs of decline, including dead branches and a bit of exposed dead wood (**b**), to extreme old age, with little surviving foliage and live bark confined to a narrow strip on one side (**c**).

a long evolutionary history; the earliest known representatives were present about four hundred million years ago. Tree-sized lycopods formed a major component of the forests that dominated the landscape of the Earth during the Carboniferous Period (about 390–300 million years before present), providing the organic material from

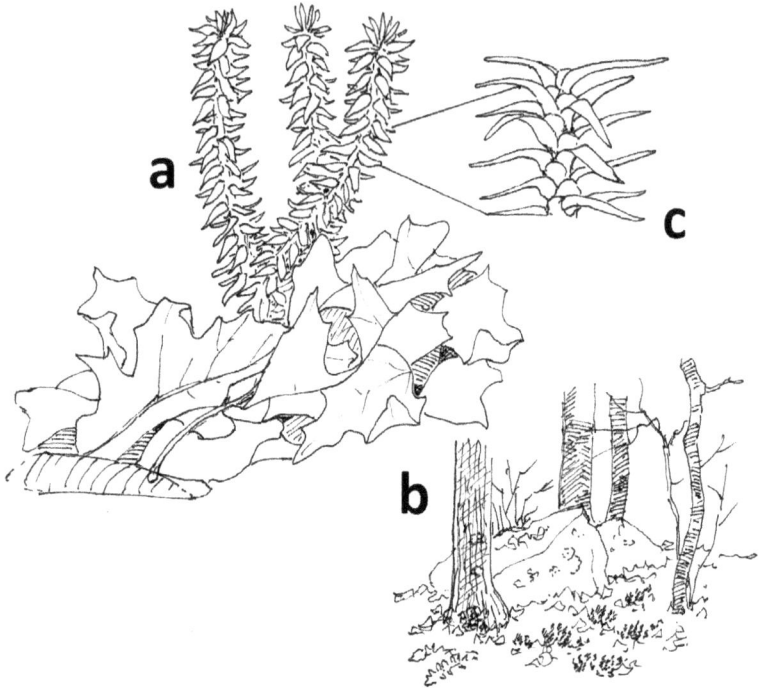

FIGURE 10.14. Shining firmoss (**a**) in a typical forest habitat (with leaf litter for scale); clumps of firmoss, a type of clubmoss, poking through leaf litter on the forest floor (**b**); and the leaf-scale attachment pattern on the stalk (**c**).

which coal was ultimately derived. You can see how the small, pointed leaves of the firmoss are attached to its stem, and similar leaf attachments on the tree lycopods of the distant past left behind a distinctive pattern of scars on the "bark" of those trees, as reflected in sandstone fossils (see Figure 2.9, c). Today, the lycopods tend to be relatively small and insignificant plants, and this is certainly true of shining firmoss. As the common name suggests, this plant bears some resemblance to a moss, but unlike a moss, firmoss has vascular tissue. The plant occurs as a series of erect stems, each about six inches high and densely covered with narrow, lance-shaped, irregularly toothed leaves. The erect stems arise at intervals from a creeping, branching underground (though often occurring just beneath the litter layer) rhizome that can be several feet in length. Firmoss is evergreen, and its bright green color is most apparent in winter. Like ferns and mosses, fir-moss reproduces by means of spores, and careful examination of the upper portion of some plants will often reveal sporangia nestled at the bases of the leaves. Because lycopods such as firmoss are evergreen, they have often been collected and used as decorations at Christmas. Unfortunately, this has led to populations in some regions being extirpated. Under no conditions should such plants be collected. Instead, they should be left in place to serve as reminders of the types of plants that existed so very long ago.

CHAPTER 11

Wildflowers, Ferns, and Other Plants of the Forest Floor

"Though April showers may come your way, they bring the flowers that bloom in May." This popular saying is undoubtedly familiar to many of our readers, and the flowers of various herbaceous plants are probably the most conspicuous element in Ozark forests during spring. In Arkansas, however, many of these "spring wildflowers" actually bloom well before the beginning of May, and a few species appear as early as late February. By early summer, the period of flowering for most plants of the forest interior has ended. During the summer months and into fall, plants that produce flowers are more likely to be found at forest margins, along roadsides, or in open fields and meadows. However, various nonflowering plants, including ferns, are present on the forest floor throughout the year. Indeed, some of these are more obvious during winter than at any other time of the year.

The plants we call "spring wildflowers" simply take advantage of the post-winter period (or "vernal season") when the forest floor first becomes relatively warm and sunny but before leaves have appeared on the trees. Once the leaves have been produced and the forest canopy closes, there is no longer enough light to sustain their growth. For many of the species in this ecological group, the aboveground (vegetative) portion of the plant doesn't persist for more than a few weeks, after which there is little or no evidence that it was ever present. However, that doesn't mean that the plant has died. Instead, most spring wildflowers are perennials, which means that they live for a number of years, surviving underground in the form of various types of specially modified stems (such as rhizomes or bulbs). From one spring to the next, such plants simply pass through a period in which they are essentially dormant. We will start with a survey of the wildflower season, tracking the seasonal sequence of flowering and growth of the most familiar wildflowers. But this is a book about forensic investigation, so we will also consider evidence related to interesting stories buried in the forestfloor (or sometimes hiding in plain sight), including some fascinating facts about how flowers attract pollinators and then disperse their seeds. Knowledge of such relationships can enhance our appreciation of wildflowers as we observe them in the forest.

The appearance of the silver-furred flower buds of hepatica in late February can be considered the first real indication that winter is beginning to give way to spring in the forests of Arkansas (Figure 11.1, f). At first, the flower buds are nodding and likely to be covered by forest-floor litter, but the distinctive three-lobed (liver-shaped) leaves, which have survived the winter, make this an easy plant to identify in the field. It is a simple matter to move the litter aside to reveal the flower buds. In a few days after their first appearance, the stalks ("scapes") of the flower buds straighten and quickly open to produce flowers that are about three-quarters of an inch across. Flower color varies from deep purple to white, although in the latter instance there is nearly always at least a trace of blue or lavender. Soon after flowering, the new leaves of the plant appear, ultimately to replace the old leaves. Both the common and genus names of this plant are derived from the Greek word for "liver" and refer to the general shape of the leaf (and another of its common names is "liverleaf"). Several species of hepatica were once recognized, but most specialists now consider the plant a single species that is distributed across North America and Eurasia. We see two distinct varieties in the Ozarks. One has sharply lobed leaves, produces flowers that are pale pink or white, and is associated with limestone ledges and outcrops. The other has rounded leaves, with flowers that vary from deep blue to pale lavender, and occurs in a wider variety of humus-rich soils. Although there is a significant overlap in the habitats of these two varieties, it is interesting that the two never seem to be found in the same location.

One of the other early-spring wildflowers, sometimes appearing in the latter part of February, is bloodroot (Figure 11.1, b). The showy white flowers, which are well over an inch across, are impos-

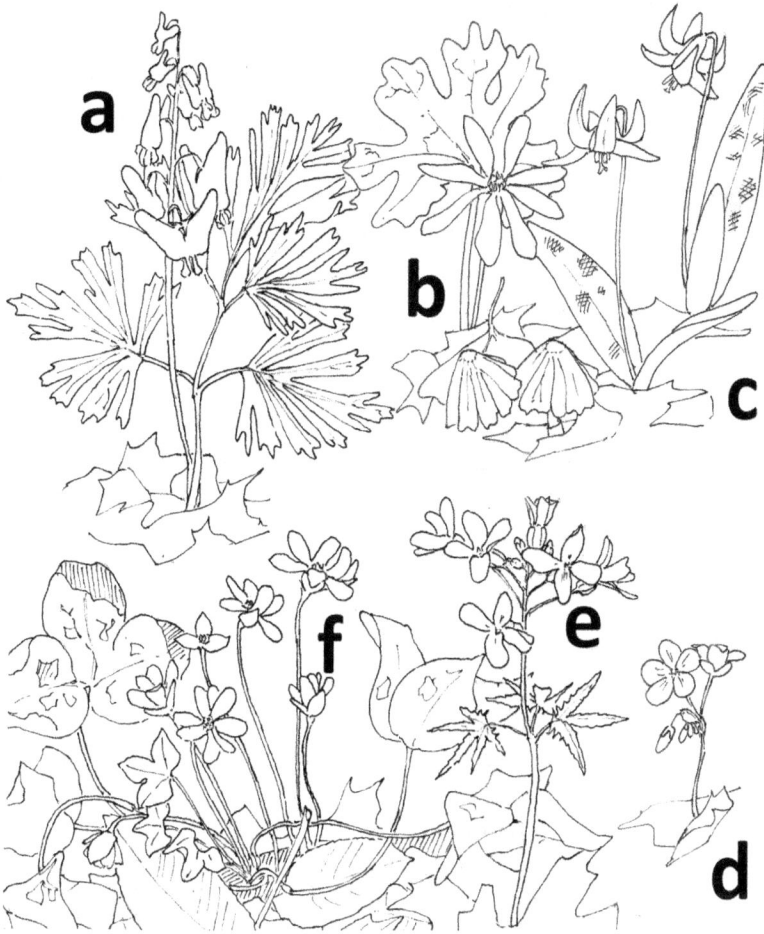

FIGURE 11.1. Some common early-spring wildflowers in the Ozarks: (**a**) dutch-man's breeches, (**b**) bloodroot, (**c**) trout lily, (**d**) spring beauty, (**e**) toothwort, and (**f**) hepatica. Young leaves of mayapple are seen emerging from leaf litter beneath the bloodroot.

sible to miss. When the plant first appears above ground, the flower stalk is more or less embraced by a relatively large-lobed leaf, which gradually expands. Each bloodroot plant has a single flower that loses its petals after a few days (or even more quickly if an effort is made to pick the flower). The common name refers to the red, blood-like fluid that oozes out when any portion of the plant, but especially the rootstock, is cut. The fluid readily stains the skin and was used by Native Americans to paint their bodies, clothing, and other objects.

The most abundant spring wildflower in many forests throughout the Ozark region is spring beauty (Figure 11.1, d), which can form an almost complete ground cover on favorable sites. Although it doesn't appear quite as early as hepatica or bloodroot, more people are likely to notice it, simply because it is so common. Moreover, spring beauty is not restricted to the forest floor and can be found in some open areas, including lawns. The plant is relatively small, usually reaching a height of no more than a few inches. Each individual bears two narrow leaves, which are often hidden in the leaf litter, and a loose cluster of no more than about ten flowers, only some of which are open at any one time. Each flower is about half an inch across, and the petals are white or pink, with distinctive, darker pink veins. Other early-spring wildflowers of note are dutchman's breeches, trout lily, and toothwort (Figure 11.1), all of which complete their life cycle in the few weeks before the forest canopy leafs out.

Another distinctive spring wildflower is Jack-in-the-pulpit, whose flowering season can extend from March through May (illustrated with several other mid-spring wildflowers in Figure 11.2, e). Most common in low, moist forests and as much as a foot and a half tall, Jack-in-the pulpit is easily recognized by its long-stalked leaves, the upper part divided into three equal parts ("leaflets"). The plant's common name is based on the unique structure of its reproductive portion. The flowers are tiny and located near the base of an elongated, column-like structure ("Jack") which is more or less surrounded by a hood-like modified leaf (the "pulpit"). In more technical terms, the column is a "spadix" and the hood-like structure is a "spathe." Both are characteristic of the plant's family (Araceae), which is largely tropical. Native Americans collected the bulb-like base of the plant for food, which accounts for another of its common names, "Indian turnip." The peppery-tasting bulbs were boiled thoroughly before being consumed, in order to dissolve the crystals of calcium oxalate in their tissue. By late summer, Jack-in-the-pulpit can be recognized by the dense cluster of bright red berries that have developed on the expanded spadix (see Figure 11.5, e).

Jack-in-the-pulpit is unusual in a number of ways that should be

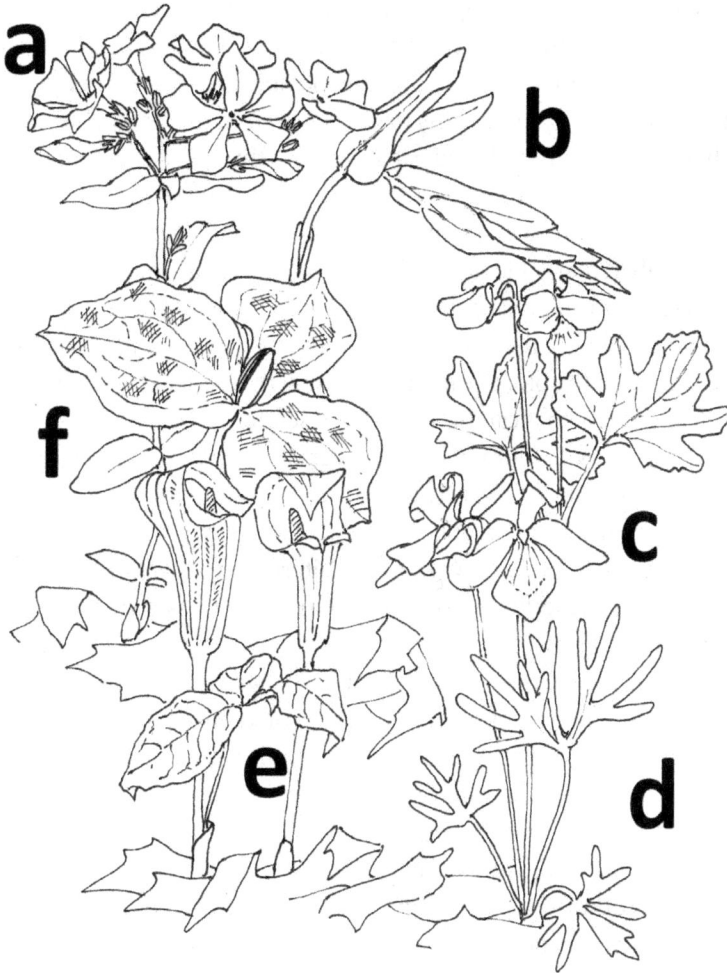

FIGURE 11.2. Some common mid-spring Ozark wildflowers: (**a**) blue phlox, (**b**) bellwort, (**c**) cut-leaf violet, (**d**) three-lobed violet, (**e**) Jack-in-the-pulpit, and (**f**) toad-shade trillium.

appreciated by any hiker out exploring the spring woods. Like many plants, it is "dioecious," which means that the male and female flower parts are on separate individuals (plants that are "monoecious" have both male and female organs on a single plant, sometimes located together in a single flower). What's unusual about Jack-in-the-pulpit is that it changes sex during the course of its lifetime. As soon as a seedling grows large enough to support a flower, it becomes male. Then, when the plant has become more robust, it converts to female. The oxalic acid crystals in the plant make it highly unpalatable for deer, but in modern forests where deer populations have increased to unsustainable levels, the hard-pressed animals will eat almost anything that is green. When mature Jack-in-the-pulpit plants are browsed by desperate deer, they are forced to revert to being male again. But the ability to change sex is not the only remarkable habit of this species. Jack-in-the-pulpit is also effectively immortal. Each growing season, the plant abandons its old corm and root system, growing an entirely new corm and set of roots from the top of the old. The plant effectively renews itself entirely every year in a process that can be repeated indefinitely.

The flowers of Solomon's seal appear somewhat later in the spring, usually during April or May (Figure 11.3, b). The arching stems of this plant are usually no more than a foot tall, though larger individuals are sometimes encountered. Two rows of alternating leaves extend along each stem, and the flower stalks arise from the bases of the leaves. The individual flowers are tubular, light yellow-green, and one-half to three-quarters of an inch in length. The stem arises from a thick, twisted, creeping underground stem (or "rhizome") that is white, with large circular scars. These characteristic scars, which have the general appearance of the wax impressions once used to seal documents—along with internal markings that have been likened to letters of the Hebrew alphabet—give Solomon's seal its name (see Figure 11.6, f). The fruit produced by Solomon's seal is a small, blackish-blue berry about the size of a pea. It is not edible to humans but is consumed by some animals. A relative of Solomon's seal, the bellwort (Figure 11.2, b), is another member of the lily family that has relatively inconspicuous yellow flowers, but it first produces a cluster of bright yellow-green leaves at the nodding end of the emerging shoot, with an entire structure that looks as attractive as many of the real spring flowers.

The vibrant yellow flowers of the trout lily also contribute to the colorful floral display on the forest floor in early spring, usually during late March and early April (Figure 11.1, c). The common name is based on the plant's mottled leaves (with purplish or whitish blotches on an otherwise green background), which are suggestive of a trout's coloring. Another common name ("fawn lily") links the same color pattern

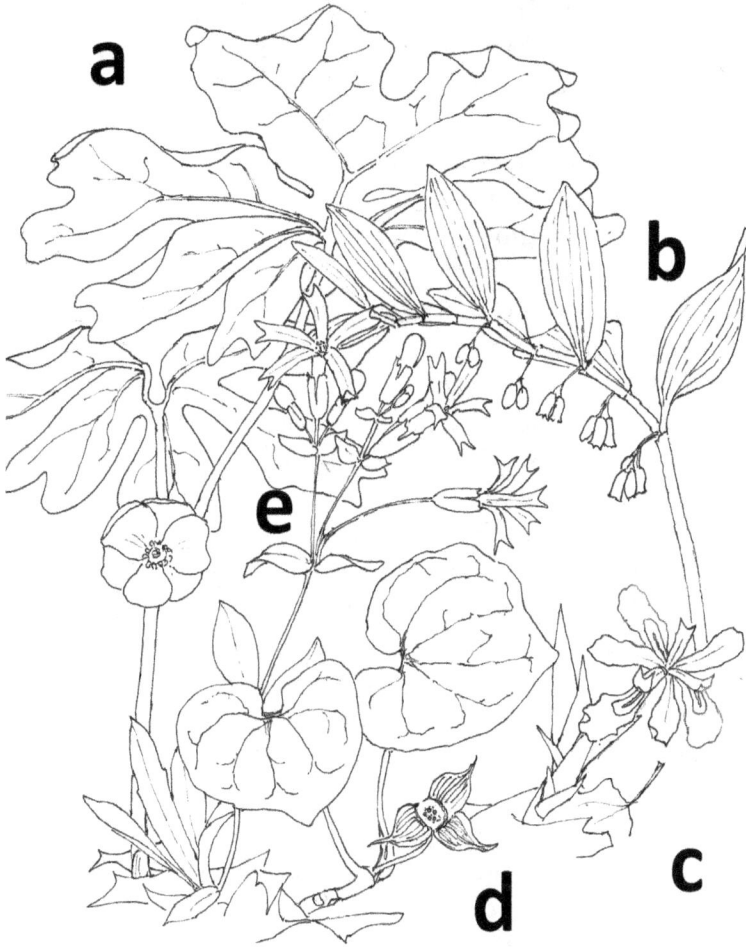

FIGURE 11.3. Some common late-spring Ozark wildflowers: (**a**) mayapple, (**b**) Solomon's seal, (**c**) crested iris, (**d**) wild ginger, and (**e**) fire pink.

to that of a fawn. The flower is about an inch and a half across and occurs about eight inches above the ground on a slender stalk that arises between two erect, elliptical leaves. Large colonies of trout lily are often found along shaded streams and on moist, forested slopes. There are two species of trout lily in the Ozarks, one bearing a yellow flower and the other a white. The yellow variety is often found in rich, moist soil in stream bottoms and regularly forms thick carpets of attractively mottled leaves, though only a small fraction of the most robust plants bloom each season. The white variety is more often found on uplands and rarely grows in dense, crowded carpets, but it makes up for its lower numbers by having almost every single plant bear a blossom. One suspects that the difference in habitat accounts for this divergence in what would seem to be closely related plants. The yellow trout lily grows in relatively well-shaded environments, where it may be more advantageous to put most resources into vegetative reproduction. By contrast, the white trout lily grows in more open upland environments, where reliance on sexual reproduction might work better in the long run.

A number of other wildflowers are found in Ozark woodlands during those delightful days of early spring when sunlight floods down through the bare branches of our mostly deciduous trees. These early flowers include dutchman's breeches, with its distinctive flowers, and toothwort (Figure 11.1, e), which can be exceedingly abundant in moist forests throughout the Ozarks. Among the more common mid-spring wildflowers are blue phlox (Figure 11.2, a); several species of bellwort (Figure 11.2, b); fire pink (Figure 11.3, e); and various violets with blue, yellow, or white flowers (Figure 11.2, c, d). Many wildflower enthusiasts think that the most spectacular of Ozark spring wildflowers is the intricate, orchid-like flower of the crested iris (Figure 11.3, c), which can spread into colonies that cover an entire hillside.

One of our best-known wildflowers of late spring is the mayapple, which is more likely to be recognized from its distinctive leaves than from its flowers (Figure 11.3, a). The deeply lobed, intricately folded, umbrella-like leaves can be seen emerging from the leaf litter shortly after the first wildflowers are in bloom (Figure 11.1, b). Although examples with a single leaf are not uncommon, only those with two leaves produce flowers. Creamy white and more than an inch across, the flowers often go unnoticed unless one takes the time to bend over and look under the relatively large leaves. Each plant produces a single flower, which arises at the junction of the two leaf stalks. Interestingly, the flowering period of mayapple occurs at the same time as the usual fruiting season of morels, highly edible fungi that are collected during spring. In any population of mayapples, it is not unusual to notice irregular yellow-orange patches on some of the leaves. These indicate the presence of a fungal parasite (a type of plant rust) that commonly infects this species.

Mayapple is one of the most widely distributed wildflowers in the Ozarks. It is seen in second-growth forests where many other wildflowers have not been able to gain a foothold. This is probably explained by the apple-like fruit, from which the plant's common name is derived (see Figure 11.7, b). One of the primary consumers of this fruit is the three-toed box turtle, a common inhabitant of our deciduous forests. Studies have shown that the passage of mayapple seeds through the turtle's digestive system enhances their germination. Box turtles can cover a considerable distance in a day, transporting mayapple seeds to new forest habitat and fertilizing them in the process.[1] Mayapple was also one of the first plants transported to Europe by early American settlers, apparently motivated by its novel form (it has no similar relatives in Europe) and the apparent edibility of the fruit, which is reputed to have a lemon-like taste. However, we strongly recommend not partaking of these "apples," because of the toxins known to be present in the leaves and roots.

On moist slopes, especially those on shaded roadsides or along streams, the dark green, velvety, heart-shaped leaves of wild ginger are a common sight in late spring (Figure 11.3, d). This plant prefers deep, rich soils, and the slopes on which it occurs often support a layer of leaf litter. The leaves (usually two per plant) arise from a long rootstock that is characterized by a ginger-like odor and taste, which accounts for the common name. Despite the fact that humans (including Native Americans) have consumed wild ginger, results obtained from modern studies reveal that the plant contains potentially carcinogenic substances and thus should be avoided. The flowers of wild ginger are not showy and are likely to be overlooked by most people,

---

1. This is not an isolated circumstance. Many plant species depend on the digestive processes of animals and birds that consume their fruits. In fact, this situation can be extreme. Biologists have found that most rainforest trees in northern Queensland, Australia, are closely dependent on seeds being passed through the digestive tract of a large, ostrich-sized bird. That bird, the cassowary, is now rare and endangered, and its elimination from the forest poses a real threat to the existence of this rainforest ecosystem.

since they are produced close to the ground and are often hidden beneath dead leaves on the forest floor (Figure 11.3, d). However, if the dead leaves are brushed away, the flowers are easy to locate. Each of the somewhat jug-shaped flowers is a dull purplish-brown. Available evidence suggests that the flowers are largely self-pollinated, but small flies that are exceedingly common in forest-floor litter may play some role. Wild ginger is one of a number of spring wildflowers whose seeds are dispersed by ants. Plants characterized by this type of dispersal (termed "myrmecochory") produce seeds with an attached food body (called an "elaiosome"; see Figure 11.7, f). Ants are attracted to the food body and carry the seed away from the plant on which it was produced. Myrmecochory is quite common and is employed by such familiar Ozark wildflowers as hepatica and trillium.

Various species of violets also flower during spring, but in these plants the vegetative portion persists throughout the summer. However, the leaves produced during spring ("spring leaves") may differ appreciably from those ("summer leaves") produced later in the year. Violets are identified by a combination of features, the most important being flower color (yellow, white, or blue), leaf shape, and the location where the leaf is attached. Leaf attachment is either "stemmed," with the leaves attached to a stem that occurs above the ground; or "stemless," the leaves appearing to arise directly from the ground, since the stem itself is underground.

The two most colorful, and most often noticed, species of violet in the Ozarks are the birdfoot violet, with its large, pansy-like flower and deeply lobed "bird's foot" leaves; and the three-lobed violet, with its relatively large, deep blue flowers and a few narrow lobes (or "cuts") in otherwise large oval leaves (Figure 11.2, c, d). The attractive flowers of violets that appear in the spring are not the only flowers produced by the plant. During the summer, violets also bear small, self-pollinating (or "cleistogamous") flowers that never open. These "summer flowers" are much smaller and very different in shape from the showy and more familiar spring flowers. Moreover, they tend to occur at or near the surface of the ground, often more or less covered by litter, and are thus not easy to spot. However, examples can be located with a little careful searching. When it comes to seed dispersal, violets don't have a specific reward for seed dispersers such as those found in many other flowers. Many violet species' seed pods open in such a way that the sides of the pods squeeze the seeds, "popping" them out for a flight of several feet.

Another common spring "wildflower" is certainly a flower but is actually most notable for its attractive leaves. This is the toad-shade trillium with its little, inconspicuous, red-brown petals hidden in the center of an umbrella composed of three wide and attractively mottled leaves (Figure 11.2, f). Like mayapple, toad-shade is one of the most widely distributed spring wildflowers in second-growth forests where other wildflowers have not yet been able to become established. Another unusual aspect of trillium (as of Jack-in-the-pulpit) is the broad shape of the leaves. These leaves resemble those of many other broadleaf plants classified as dicots (plants with seed leaves or cotyledons that appear in pairs), but trillium is actually a monocot (plants like grass and lily that produce a single cotyledon).

The plants described above are only a few of the spring wildflowers found in the forests of the Ozarks during late spring. Among the other species likely to be encountered are toothwort, false Solomon's seal, purple trillium, fire pink, and wild geranium. Not all of these produce their flowers at the same time in the spring or occur in the same types of ecological situations, so it is necessary to make more than a single trip into the field to observe them. In fact, the flowering season often extends from the last week in February until the middle of May, a period of almost three months.

As already noted, relatively few plants produce flowers in the forest interior after the canopy has closed. Moreover, one of the more common examples of those few flowers that do bloom in the deep forest in midsummer might not even be recognized as a flowering plant. This plant has been called by various names, including "ghost pipe," "corpse plant," and "ghost flower," but the most widely used common name is "Indian pipe" (Figure 11.4, e). This plant has a white, waxy appearance and usually occurs in the deep shade of the forest interior during mid- to late summer. Because of its general appearance, Indian pipe is occasionally mistaken for some type of fungus. However, it is actually a very unusual vascular plant. When Indian pipe first emerges from the litter layer on the forest floor, the top of the plant is curved downward, so that the single flower (the specific epithet, *uniflora*, means "one-flowered") faces the ground. The resemblance to a small, slender pipe accounts for its most widely used common name. Later, after the flower has been pollinated, the stem grows in such a way that the flower turns upward. At this time, the entire plant darkens to a color ranging from pale pink to sometimes a deep red. Eventually, after the stem dies, it becomes black. Indian pipe

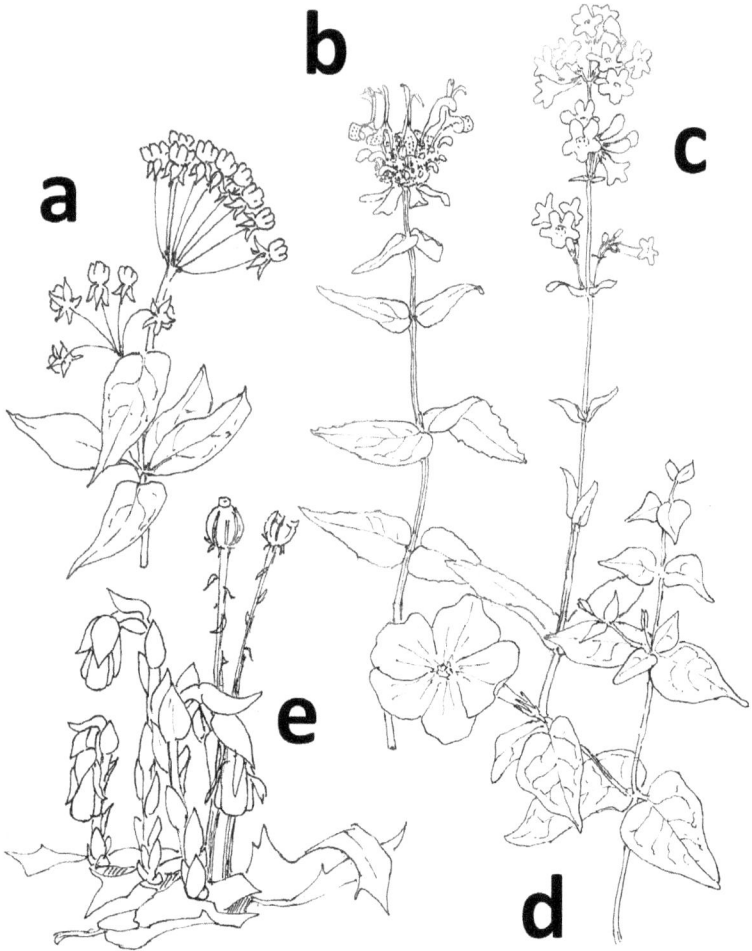

FIGURE 11.4. Some common early- to mid-summer Ozark wildflowers: (**a**) four-leaved milkweed, (**b**) horsemint, (**c**) foxglove penstemon, (**d**) smooth or "wild" petunia, and (**e**) Indian pipe (with remnants of the previous season's seed pods).

is what is known as a "mycoheterotroph." This means that the plant depends on a nearby tree (usually an oak) to meet its energy needs. That energy is transferred from the tree to the plant through a shared network involving an ectomycorrhizal fungus (to be discussed in the next chapter). There is no direct connection between the Indian pipe and the tree. Instead, the two are linked through the fungus.

Other flowering plants that can be found in the forest interior in the summer include downy rattlesnake plantain (actually a type of orchid; see Chapter 10) as well as four-leaved milkweed, horsemint, foxglove penstemon, and smooth petunia (Figure 11.4). The first two of these species produce flowers that are not particularly colorful, but at this time of year most of the plants in flower are outside rather than inside the forest, so the bright blue flowers of the wild petunia are especially appreciated in the woodland interior.

With the arrival of summer, members of one family of flowering plants, the Asteraceae, become increasingly abundant and conspicuous. This family includes the asters and goldenrods, along with such other Ozark plants as wild chicory, ironweed, and ox-eye daisy. The latter is especially familiar to most people and often occurs in large populations in open fields and along roadsides from mid- to late summer (Figure 11.5, b). Ox-eye daisy is not native to North America and was introduced from Europe as an ornamental plant in the nineteenth century. Since then, it has become firmly established throughout much of the continent. As such, it serves as an excellent example of an organism that has greatly expanded its natural range as a result of human activities. The flowering stems can reach a height of more than two feet, and the flower head (what appears to be a single flower in the Asteraceae is actually a composite structure consisting of numerous small flowers) is about two inches across, with white, petal-like ray flowers surrounding a bright yellow, dome-like central structure made up of disk flowers.

While members of the Asteraceae tend to be characteristic of open, generally drier sites, there are other flowering plants that are typically restricted to places where there is abundant moisture. One such plant is the orange jewelweed, which is commonly associated with low, wet areas, often growing along streams and in roadside ditches (Figure 11.5, d). The golden-orange flowers with scattered reddish-brown splotches are distinctly three-lobed, and the rear of

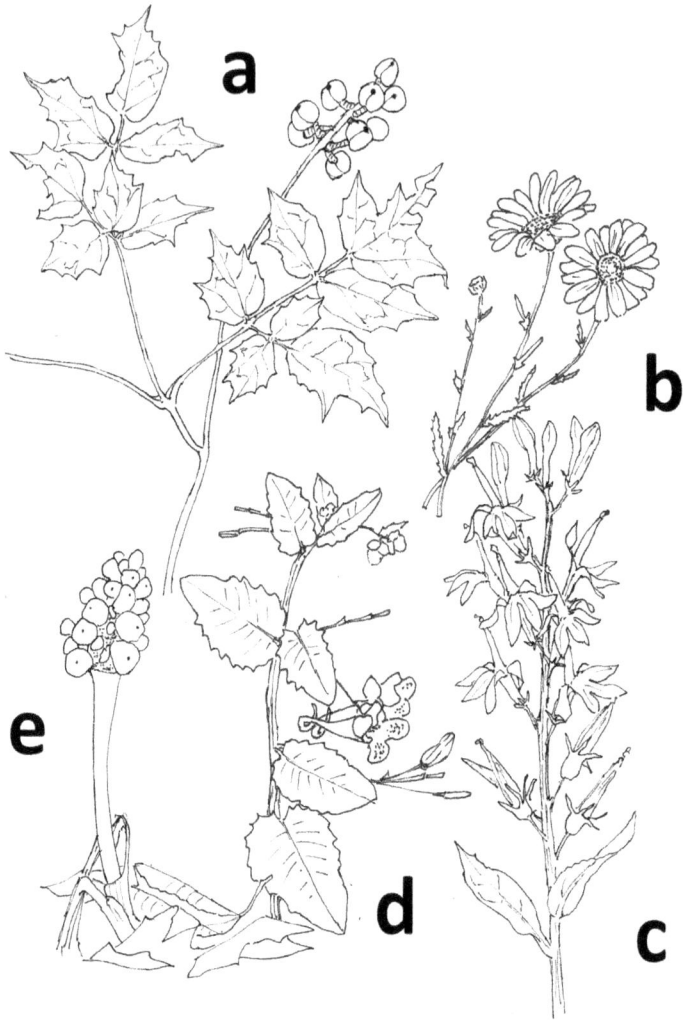

FIGURE 11.5. Some common late-summer and autumn Ozark wildflowers, along with the attractively colored fruits of other flowers that have produced blossoms earlier in the year: (**a**) baneberry (also known as "doll's eyes"), (**b**) ox-eye daisy, (**c**) cardinal flower, (**d**) jewelweed, and (**e**) Jack-in-the-pulpit berries.

each flower extends backward to form a distinctive hooked spur. The stems of jewelweed are somewhat translucent and succulent, and when held underwater they have a silvery ("jeweled") appearance, which is thought to have given rise to the common name. The mature seed pods produced from the flowers have projectile seeds that explode out of the pods when the latter are touched, which accounts for another common name ("touch-me-not"). Jewelweed has been considered a traditional remedy for skin rashes, including that caused by poison ivy. However, modern clinical studies have not provided evidence to substantiate this use.

Perhaps the most spectacular flowers of late summer are those produced by the cardinal flower (Figure 11.5, c). The plant is up to four feet tall and occurs in wet places, typically along stream banks or at the edges of ponds. The flowers, produced in an elongated cluster at the apex of the plant, are deeply five-lobed, more than an inch and a half across, and a brilliant red (the common name refers to the similar color worn by Roman Catholic cardinals).

Another member of the same genus is the great blue lobelia (Chapter 4), which can be found in similar habitats but doesn't grow quite as tall and has flowers that are less deeply lobed and violet-blue in color. The plant was once reputed to cure syphilis (hence its specific epithet, *siphilitica*). Interestingly, although cardinal flower and great blue lobelia are very closely related, they are pollinated by two entirely different groups of organisms. The violet-blue flowers of great blue lobelia are pollinated by bees, while the red cardinal flowers are typically pollinated by hummingbirds. In both color (red is not very visible to insects) and structure, the latter species' flowers are adapted to pollination by hummingbirds, which are equipped with beaks and tongues that can reach the nectar contained within the deep throats of cardinal flower's trumpet-shaped blossoms. Columbine and fire pink (Figure 11.3, e), which also have deep, tubular flowers with a bright red color, are two other examples of plants usually visited by hummingbirds.

The fall flowering season begins in September and extends to the beginning of winter, which, in some particularly mild years, can be as late as mid-November. Plants producing flowers during this period include boneset, wingstem, joe-pye weed, daisy fleabane, wild carrot, and various sunflowers. With the exception of wild carrot, these are all members of the Asteraceae. There are also a number of plants that have attractively colored fruit that catch one's attention at this time of year. These include the brilliant red berries of ginseng (see Figure 10.3, a) and Jack-in-the-pulpit (Figure 11.5, e) and the starkly white fruits of the baneberry, offset by thick reddish stems and a small black "pupil" that gives that plant the common name "doll's eyes" (Figure

11.5, a). Not to be forgotten in this context is the common dandelion, whose flowers are sometimes possible to find during every month of the year. The dandelion is yet another example of an introduced plant (from Eurasia) that has become firmly established in North America.

One aspect of wildflowers in general is that their attractive blooms were not created for our enjoyment. Flowers are an important part of the reproductive process and are mostly designed for the transfer of genes to offspring. As noted above, close attention to the shape and structure of flowers can tell a lot about the specific pollinators that do the hard work of transferring pollen between those flowers. Both pollen itself and a secreted sugar-rich fluid (nectar) serve as motivation for the insects to visit flowers. Some flowers, such as the trout lily, are generalists, providing nectar and pollen to a wide variety of insects. The trout lily increases the likelihood that a bee will transfer pollen to another flower by having two sets of anthers (pollen-bearing organs) in the flower that mature in sequence, thus doubling the period in which bees are exposed to pollen in the flower. Other plants have more specific pollinators in mind. Some flowers with deep throats rely on long-tongued bumblebees. A few plants, such as the Jack-in-the-pulpit, have slightly fetid odors that attract gnats and flies. The bright colors we enjoy on some of our favorite flowers are there to attract the attention of pollinators.

Interestingly, it has been found that widespread flowers such as hepatica tend to be more brightly colored in the western part of their range (including the Ozarks), while they are more likely to be pale or white in the east. This is sometimes credited to the relatively more dense forests of the eastern United States, where white flowers provide more of a contrast in the shaded woodlands than they do in the more open forests of the Ozark Plateau. Other flowers have petal shapes and arrangements designed to force pollinators to douse themselves with pollen as they visit each blossom. Sometimes, stamens shed pollen before ovaries are receptive, thus ensuring that cross-pollination occurs. In other cases, the ovaries become receptive while anthers are still shedding pollen, to allow for self-fertilization as a default option if cross-pollination has not already occurred. The variety of the flowers we see throughout the year results from the complexity of this evolving interaction between plants, their pollinators, and the environment around them.

Although humans most often focus on the showy floral facet of the plant reproductive process, other reproductive processes are also underway on and under the forest floor. We have already seen how the cleistogamous flowers of violets disseminate seeds without the benefit of pollination, in a way that is hardly visible above the leaf litter. Another interesting example of vegetative reproduction produces the carpets of leaves of the yellow trout lily found in moist bottomland soils; this is accomplished by lateral shoots that arise from the bulblet located at the base of the plant (Figure 11.6, c). This shoot or "dropper" (the technical term is "stolon") arches outward and embeds itself nearby to form a new bulblet at its tip. Thus, the entire carpet of trout lily leaves you see likely represents a single plant (a clone). Many other flowers have means of vegetative propagation through underground stems, corms, and bulbs (Figure 11.6). Patches of crested iris or mayapple often can be more than ten feet in diameter as a colony expands outward from an initial colonizer. One of us had a direct quantitative experience of such wildflower expansion when three small bloodroot corms purchased from a mail-order nursery multiplied into more than fifty individual flowering plants five years after being set out in an Arkansas rock garden.

Another aspect of wildflowers in the Ozarks is the effect of human activities on the forest. Some flowers, such as hepatica, bloodroot, and wild ginger, are very slow to reestablish themselves in woodlands where they had once been eliminated when those lands were transformed into cropland or pasture. Ginseng, in particular (Figure 11.7, a), is an indicator of pristine old-growth forest because it is slow to spread into second-growth forest and has been extensively harvested in the wild for its reputed value as a medicinal plant (Chapter 10). As shown in a previous chapter, tree-ring studies demonstrate that old-growth forests do not necessarily have to be characterized by large-diameter trees, so the presence of flowers such as bloodroot and hepatica can help identify small plots of undisturbed growth that have been hiding in plain sight. The presence of relatively uncommon wildflowers and the general amount of species diversity can indicate the degree of habitat disturbance in the past. The human impact on the landscape can also be manifested in less obvious ways. One such impact is the elimination of large predators, allowing dense populations of deer to eliminate or severely reduce the abundance of the more palatable

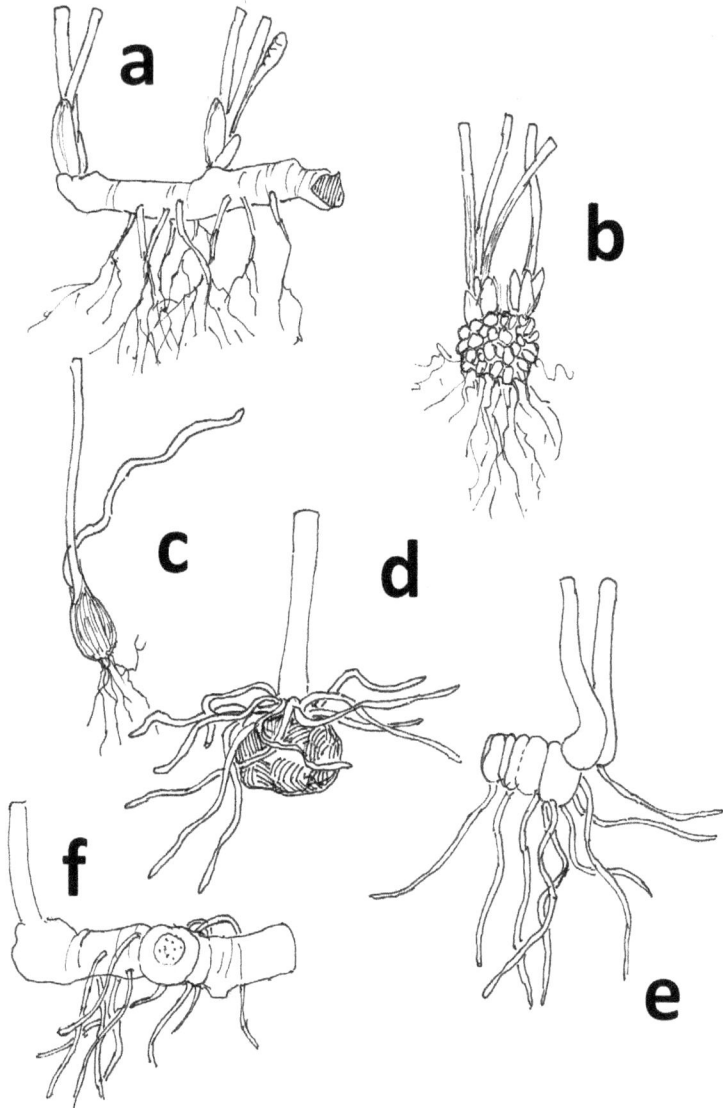

FIGURE 11.6. Examples of the underground structures (bulblets, corms, and rhizomes) of some species of Ozark wildflowers: (**a**) bloodroot, (**b**) dutchman's breeches, (**c**) trout lily shown with a typical developing stolon, (**d**) Jack-in-the-pulpit, (**e**) trillium, and (**f**) Solomon's seal (note the "seal" that gives this plant its name).

herbaceous species of plants. As such, deer are every bit as important in the reproductive ecology of wildflowers and forest herbs as they are in the establishment of overstory species. As already noted, such unpalatable species as Jack-in-the-pulpit can be affected in subtle ways. Another subtle but pervasive human-introduced change is the addition of Eurasian earthworms into the soil development process. Earthworms divert the decay of organic matter by natural fungal organisms to a process more dependent on bacteria. This represents a major change in soil structure that continues to have important implications for the future of our forests.

All the plants mentioned thus far in this chapter are vascular plants (i.e., they have water-conducting tissues in their roots, stems, and leaves), and all of them reproduce by producing seeds. However, there are some types of vascular plants found in the forests of the Ozarks that do not produce seeds. Instead, these seedless vascular plants reproduce by producing microscopic spores as part of a surprisingly complex life cycle. The best known and most common examples of seedless vascular plants are the ferns (Figure 11.8). Ferns are familiar plants to most people, and few other plants can rival their graceful beauty in a woodland setting. Most species of ferns occur in shaded situations, but a few species (referred to as "sun ferns") can grow in full sunlight. One of these is the bracken fern, whose distinctive three-divided "fronds" (the term used for the leaves of ferns) are found at forest edges, old fields, and other open areas. Bracken fern tips the balance in its favor by producing toxins that discourage the establishment of tree seedlings that might encroach on its sun-drenched domain. Ferns also differ in whether they are evergreen or have fronds that die back during the fall. Many species belong to the latter category, but not the Christmas fern, which remains green throughout the winter and has been used in Christmas decorations (hence its common name). Christmas fern occurs in virtually every imaginable ecological situation and is undoubtedly the single most widely distributed fern in the Ozarks.

Most of the ferns found here occur on the forest floor, but a few species are confined to rocky areas, where they grow out of fissures and crevices on cliff faces, rock ledges, and boulders. Certain species are associated only with acidic rocks such as sandstones, while others

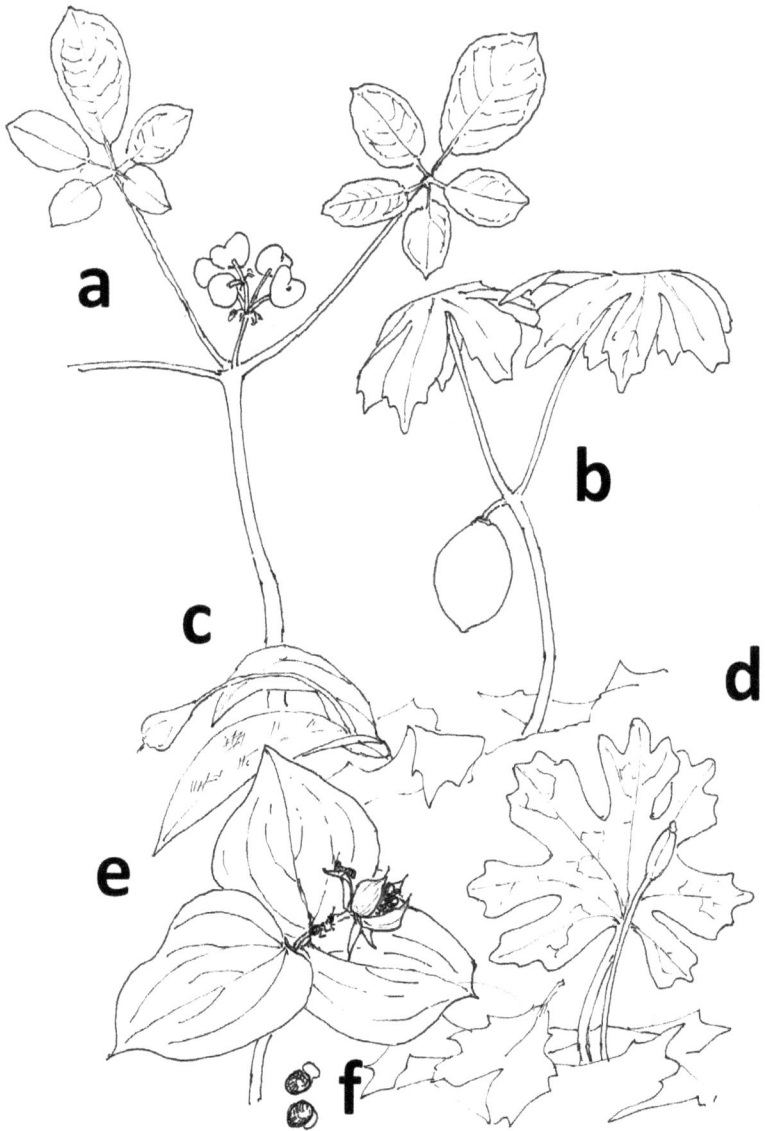

FIGURE 11.7. Examples of the seed pods produced by some typical species of Ozark wildflowers: (**a**) ginseng, (**b**) mayapple, (**c**) trout lily, (**d**) bloodroot, and (**e**) trillium—note ants present on the latter and (**f**) detail of seeds with attached elaiosomes.

are characteristically found on calcareous rocks such as limestones. For example, the walking fern's usual habitat is a moist, moss-covered limestone boulder or rocky ledge. A relatively small evergreen fern with narrow, undivided fronds that taper gradually to an elongated tip, the walking fern can reproduce vegetatively when the tip of a frond takes root, giving rise to a new plant. In this way, the fern "walks" across the boulder. Another member of the same genus, the ebony spleenwort, is a small evergreen fern characteristic of drier situations such as rock outcrops and open, often rocky forests. The ebony spleenwort has a divided frond and a smooth, glossy, red-brown leaf stalk.

Another interesting fern is the polypody, a small creeping fern that often occurs perched on the branches of oaks and other thick-barked trees where the bark can serve as a suitable substrate. It sometimes grows in thick beds of moss and leaf litter on top of great sandstone boulders in Ozark ravines and canyons (hence the common name "rockcap fern"). When growing on tree branches, the plant is termed an "epiphyte" (Figure 11.8). Vascular plants that grow as epiphytes, though rather rare in the forests of the Ozarks, are a common feature of moist tropical forests. Most of the other epiphytes present in Ozark forests are mosses and lichens, which are far less conspicuous than this charming little fern. The habitat in which the polypody grows is subject to long periods of desiccation during times of drought, such that its foliage can appear shriveled and dry, seemingly lifeless. But if you pluck a small section of such a dried-out polypody and soak it in a cup of water, you will see the fern leaves almost immediately appear to come back from the dead.

In addition to the vascular plants mentioned thus far, members of one group of nonvascular plants, the bryophytes, are well represented in the forests of the Ozarks (Figure 11.9). Although present throughout the year, bryophytes are most apparent during the winter months, when deciduous trees are bare of leaves and most other plants have died back. At this time of year, the patches of green found on decaying logs and stumps, on rock outcrops, in small streams, and girdling the bases of many trees contrast with the otherwise drab forest landscape. Bryophytes play a significant but often overlooked role in the ecological processes of forests, and various aquatic bryophytes are the dominant primary producers in small, headwater streams, which are

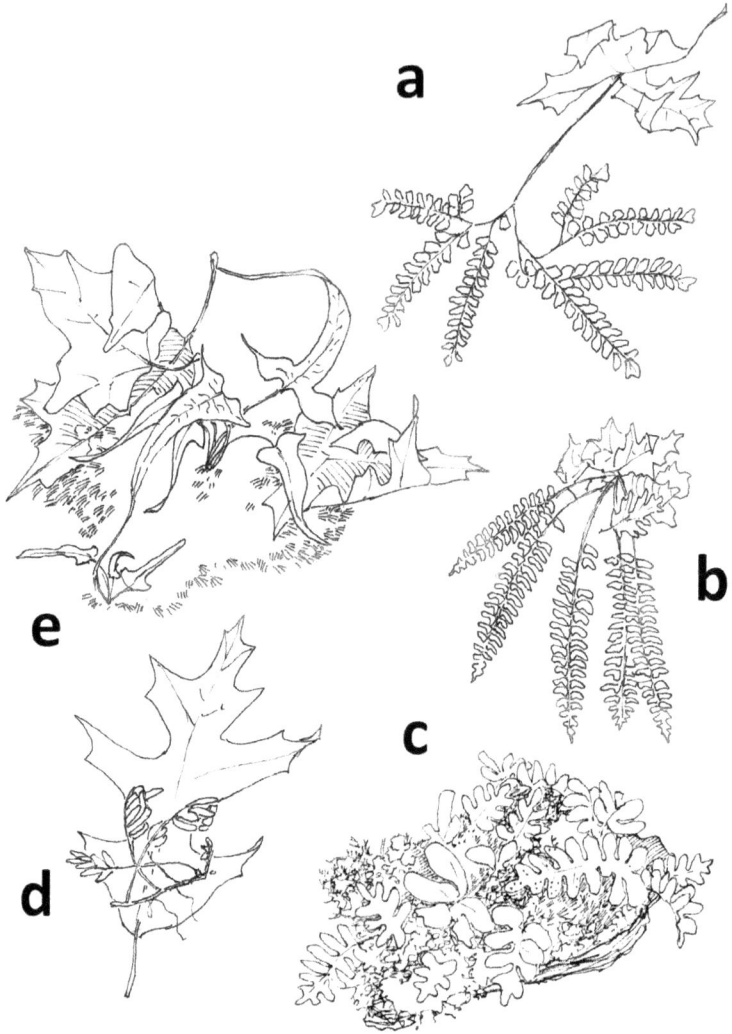

FIGURE 11.8. Some common species of Ozark ferns: (**a**) maidenhair fern; (**b**) Christmas fern; (**c**) polypody fern (shown embedded in bark from an overhead tree where it was growing as an epiphyte), with (**d**) detail showing the shriveled and apparently lifeless appearance of polypody fronds when desiccated; and (**e**) walking fern. Scale varies, as indicated by the relative size of leaf litter with each.

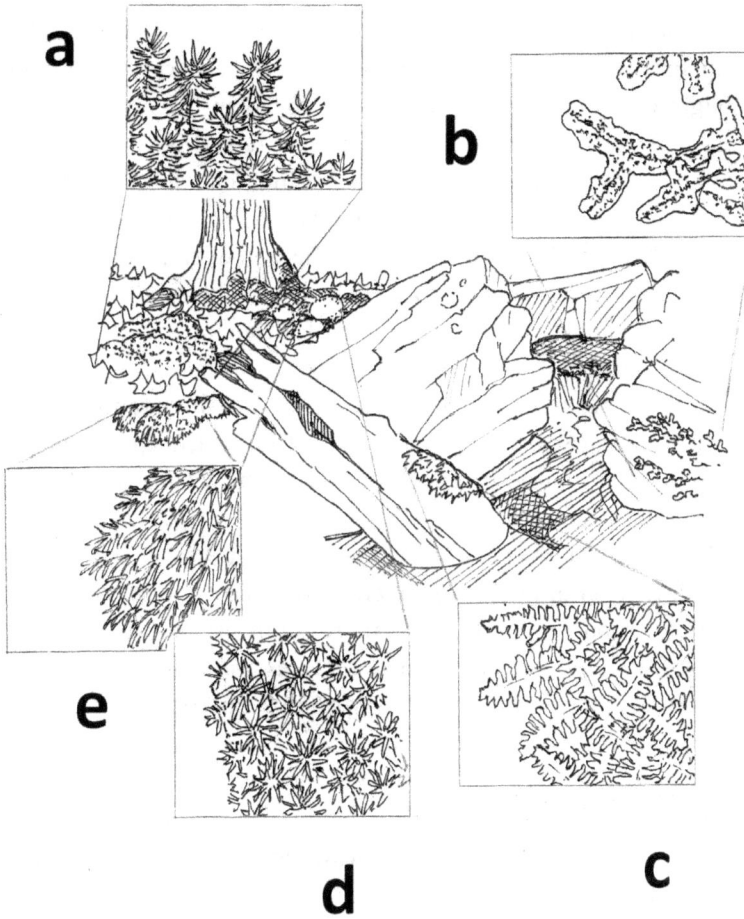

FIGURE 11.9. Some common species of Ozark bryophytes, often found in moist rock-ledge or streamside habitats like the one shown here: (**a**) haircap moss, (**b**) liverwort, (**c**) fern moss, (**d**) white cushion moss, and (**e**) broom moss.

often virtually free of vascular plants. Bryophytes are small plants, but numerous individual plants may occur together to form a loose layer or dense mat over a considerable portion of the surface of a rock or decaying log.

There are three groups of bryophytes: hornworts, liverworts, and mosses. Although members of the latter group are by far the most familiar to most people, liverworts are relatively more common in some microhabitats, such as the bark surface of living trees. As a group, bryophytes are found primarily in moist places, often under low light conditions. Many species can thrive in the full shade of a forest canopy. Bryophytes require moisture to survive because their plant body lacks the protective waxy covering ("cuticle") that prevents water loss in vascular plants. However, they do not require a constant supply of moisture and can manage quite well during periods when water is not readily available. Most people have heard the old saying that "mosses grow on the north side of a tree" and thus provide a kind of natural compass. There is some basis for this saying, since the south-facing side of the tree is warmer and drier (and thus less favorable for mosses) than the north side. However, deep inside a forest, where sunlight doesn't penetrate, mosses grow equally well on all sides of the tree trunk.

Although this discussion of the bryophyte world is necessarily brief, we encourage Ozark hikers to note how much mosses and lichens contribute to some of the more subtle aspects of the natural scenery. Stop to notice the various textures and colors of the moss cushions on ledges and boulders. Figure 11.9 can only provide a hint of the great diversity of species you can find in our local forests. Take a close look at the way mosses cling to the trunks of living trees and drape over the decaying wood of fallen logs. Such careful inspection of nature's intricate detail can add a whole new dimension to your outdoor experience.

CHAPTER 12

Mushrooms and Other Fungi

Even a relatively young child soon becomes aware that there are two fundamentally different groups of living organisms—plants and animals. However, an observant individual who walks through any forest in Arkansas cannot help but notice evidence of a third group, the members of which are just as important as plants or animals but far less conspicuous. As a result of their size (at least in most instances), plants are impossible to overlook, and the same is true of animals, whose movement and color (especially in birds, for example) tend to make them readily apparent in nature. However, the same cannot be said for most of the organisms that belong to the third group, the fungi (singular: fungus). Most visitors to the Ozark forests are probably unaware of the complex fungal world that surrounds them as they walk along a typical Ozark mountain trail.

Although they are similar to plants in some ways, fungi lack the green pigment chlorophyll and so cannot produce their own food through photosynthesis, using sunlight, water, and carbon dioxide. Instead, they meet their nutritional needs by breaking down dead organic matter or, in some cases, by attacking and living on or within plants, animals, or other fungi. Fungi that depend on dead organic matter are called "saprophytes." Fungi that feed on living organisms (their "hosts") are called "parasites" if the host remains alive and "pathogens" if the host is killed. The distinction between parasites and pathogens is not always clearly drawn. Some fungi can be a parasite of a particular host (sometimes for an extended time) and then switch to become a pathogen of the same host, resulting in the death of the latter. As such, when the casual forest hiker notices interesting-looking "mushrooms" on the forest floor or attached to the wood of a decaying log or stump, he or she may not recognize these as an indication of an entire biological assemblage hidden away within the surrounding ecosystem.

Certain other fungi form symbiotic relationships with the roots of trees and other plants. This relationship, called a "mycorrhizal association," is mutually beneficial to both the plant and the fungus. The fungus enables the plant to take up nutrients that would otherwise be unavailable, and the plant provides nutrition for the fungus. It is now known that the vast majority of plants form mycorrhizal associations. In some instances, such as in orchids (Chapter 10), this is so essential to the plant that it could not survive without its fungal partner.

The overwhelming majority of fungi are microscopic during their entire life cycle, and so poorly known that mycologists (the scientists who study fungi) have yet to determine the number of species worldwide, although the total certainly exceeds several million. The basic structure of a fungus is quite different from that of the more familiar plants or animals. Except for some of the very simplest examples, such as yeasts, in which the vegetative body is essentially unicellular, a fungus consists of a system of very finely branched, microscopic, thread-like structures called "hyphae" (singular: hypha). The entire mass of hyphae making up a particular fungus is called the "mycelium" (plural: mycelia). The mycelium typically occurs in soil, leaf litter, or decaying wood, where the individual hyphae obtain the nutrients and water the fungus needs to grow. Figure 12.1 shows the relationship between a mycelium and the fruiting body that arises from it. Because the hyphae are so small and occur within the soil, in leaf litter, or in decaying wood, they are effectively hidden from view unless one takes the time to look very closely. However, careful examination of a handful of dead leaves taken from the litter layer on the forest floor will invariably reveal the presence of numerous hyphae that appear to literally "stitch together" partially decomposed dead leaves. There are two major taxonomic groups of fungi—the ascomycetes and basidiomycetes—that do not remain microscopic for their entire life cycle. In these fungi, the mycelium gives rise to one or more spore-producing reproductive structures called "fruiting bodies." This happens only after the mycelium has undergone a period of growth, and only under favorable conditions of temperature and moisture. A fruiting body is somewhat analogous to an apple on an apple tree—it is the "fruit" of the mycelium. Most fruiting bodies last only a few days, but a mycelium may live for years.

FIGURE 12.1. Two typical Ozark fungi, morel (**a**) and agaric (**b**), each showing
the aboveground fruiting body attached to the underground mass of thread-like
hyphae (referred to collectively as "mycelium").

The mushrooms that one encounters in situations ranging from an ordinary lawn to various types of forest are the most familiar examples of fruiting bodies. Some apply the term "mushroom" to any type of fruiting body that is large enough to be seen with the naked eye. Others restrict the use of the term to those fruiting bodies that consist of an expanded, often inverted, bowl-shaped structure ("cap") that is held aloft by a stem-like structure ("stipe"). In the classical concept (which we use here), thin blade-like structures ("gills") radiate outward from the stalk on the underside of the cap. An example of this type of fruiting body can be observed in Figure 12.1, b. Because of the presence of the gills, mushrooms are often referred to as "gilled fungi," although a mycologist is more likely to use the more technical term "agaric." It should be pointed out that although the term "toadstool" is sometimes given to what are considered poisonous mushrooms, this term has no precise meaning.

Mushrooms are basidiomycetes, a group that contains most of the fungi that produce fruiting bodies large enough to be observed easily in nature. Only a few ascomycetes produce fruiting bodies this large, but some of these (the morels, genus *Morchella*) are among the best-known edible fungi. Unlike most basidiomycetes, whose fruiting bodies tend to occur during the summer and fall, morels appear in the spring (Figure 12.1, a). Although mushrooms are the examples of basidiomycetes most likely to be familiar to the average person, the group also contains numerous other forms that are distinguished from one another on the basis of their overall shape and, more importantly, where the spores are produced. Among the more commonly encountered non-gilled basidiomycetes are the boletes, puffballs, chanterelles, coral fungi, stinkhorns, thelephores, and bird's nest fungi (Figure 12.2). In some of these, including coral fungi, chanterelles, and thelephores, the spores are produced on a portion of the external surface of the fruiting body. In others, including boletes (Figure 12.2, a) and polypores (Figure 12.4), they are produced within a series of tubes that extend back into the fruiting body, in a mass of foul-smelling slime (as in stinkhorns) that covers the upper portion of the fruiting body or are completely enclosed (as in puffballs and bird's nest fungi) within the fruiting body. Regardless of where they are produced, spores of basidiomycetes have one thing in common—they develop

on the outside of a special, club-shaped hypha called a "basidium."
Each basidium typically gives rise to four spores (or "basidiospores,"
to use their more technical name). Even in a relatively small fruiting
body, the total number of spores produced is so large that it is difficult
to comprehend. Some large puffballs have been estimated to produce
as many as several trillion spores!

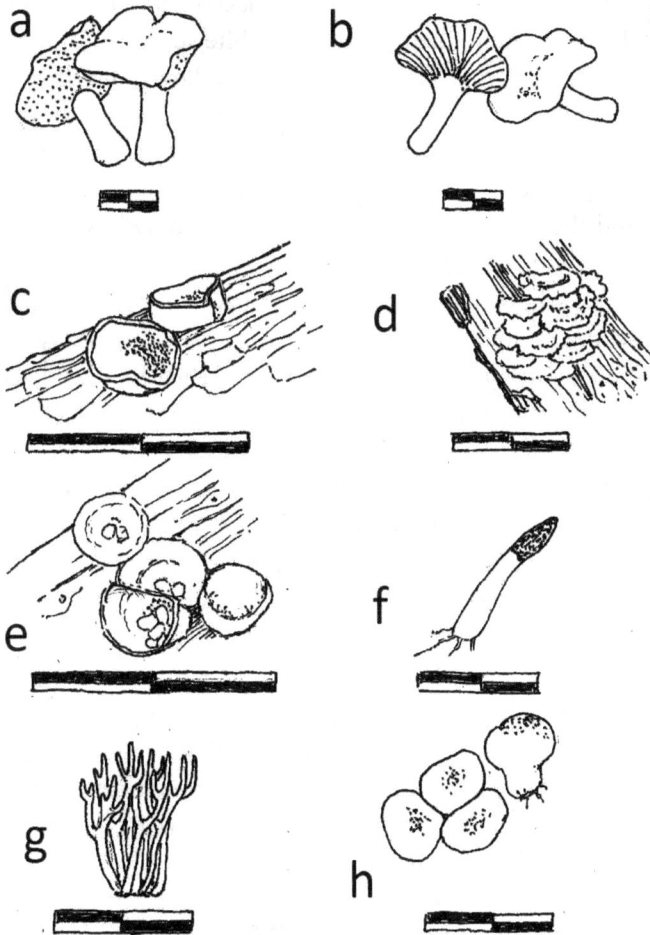

FIGURE 12.2. Some of the more common types of fungi in Ozark forests (each scale
bar represents one inch): (**a**) boletes, (**b**) chanterelles, (**c**) cup fungi, (**d**) thelephores,
(**e**) bird's nest fungi, (**f**) stinkhorn, (**g**) coral fungus, and (**h**) puffballs.

Although individual spores are too small to be seen with the naked eye, it is possible to make what is known as a "spore print" to observe the spores in mass, which allows their color to be determined. Spore color is an important feature when identifying fungi, especially the gilled fungi or boletes. A spore print is easy to make. One simply cuts off the cap at the top of the stalk and places the cap, lower surface downward, on a piece of white paper for several hours. To retain the spores and prevent excessive drying, the cap and paper are covered with a drinking glass, bowl, or dish (or placed inside a small zip-lock bag). Many common gilled fungi have white spores, but others have spore colors ranging all the way from faint yellow to black.

In ascomycetes, the second major taxonomic group of fungi, the spores ("ascospores") develop inside a special elongated hypha called an "ascus" (plural: asci). There are generally eight spores produced per ascus, and they are lined up like peas in a pod. Other ascomycetes besides morels produce a fruiting body large enough to be detected with the naked eye, although it may be necessary to do some careful searching of suitable places to turn up examples. Many of these fruiting bodies are shaped like a cup or bowl (Figure 12.2, c), with the spore-producing hyphae located on the upper surface. In fact, there are so many examples in which the fruiting body is cup- or bowl-shaped that ascomycetes as a whole are often referred to as the "cup fungi."

Fungi play a major role in maintaining the balance of nature. For example, it has been estimated that over the course of a single year, several million leaves fall to the ground in an acre of forest. These leaves don't continue piling up year after year because they are broken down by various saprophytic fungi. As a result, essential nutrients in the leaves are recycled to the soil. The fungi that decompose leaves tend to be less conspicuous than those that decompose wood, although members of the genus *Marasmius* (sometimes referred to as "horsehair mushrooms") are not uncommon if one takes the time to survey an area of the forest floor (Figure 12.3).

Fungi are the major group of organisms responsible for the decomposition of woody debris, and since many of the wood-decay fungi produce fruiting bodies that are perennial, it is not difficult to locate them in nature. In fact, examination of virtually any stump or decaying log on the forest floor, or any still standing but dead tree, is

FIGURE 12.3. Horsehair mushrooms in their usual habitat of decaying leaves on the forest floor.

likely to turn up several different examples. Many of these are poly-pores, commonly referred to as "shelf" or "bracket" fungi because their somewhat shelf-like or bracket-like fruiting bodies extend out-ward from the log, stump, or tree they are decomposing (Figure 12.4). Others belong to another group of basidiomycetes known as "thele-phores." The fruiting body of a thelephore often has the same basic shape as a polypore but tends to be smaller, thinner, and more likely to occur in overlapping clusters. More importantly, the lower surface of a thelephore is smooth, with no evidence of the small pores that are the basis of the very name "polypore." It is very easy to distinguish between a polypore (with small pores present) and a thelephore (with no evidence of pores) by examining the lower surface of a fruiting body with a hand lens.

The fruiting bodies of most polypores are tough, corky, or leathery, but there are some noteworthy exceptions. One of these is the "sulphur shelf" polypore, the fruiting bodies of which occur in overlapping, shelf-like clusters on decaying wood or sometimes still-living trees during the summer and fall. The fruiting bodies are various shades of bright sulfur yellow to orange-yellow on the upper surface and yellow on the lower surface. The young fruiting bodies are relatively soft and fleshy (especially the portion away from the base) and can be collected, cooked, and then consumed by humans. The cooked fungus

FIGURE 12.4. Polypores ("bracket fungi") growing in their usual standing-dead-tree habitat, with adjacent older bracket fungi still largely intact on much more heavily decayed wood nearby. The fungus in both situations is the common "artist's conk."

is reputed to taste somewhat like chicken, which accounts for the common name "chicken-of-the-woods" (Figure 12.5).

An example of a tough polypore found in similar situations is the fungus known as the "artist's fungus" or "artist's conk" ("conk" is another common name that has been applied to some bracket fungi). This polypore is collected in summer and early fall by artists who use a sharp instrument to scratch intricate designs on the lower surface. Examples of their work are often seen at craft fairs. The lower surface is essentially white when fresh but turns brown with age. Because fruiting bodies of the "artist's fungus" commonly reach or exceed a width of six inches, they are hard to overlook on the standing dead trees or fallen logs on which they occur (Figure 12.4).

Interestingly, as a log on the forest floor decomposes, a process that may take as much as several decades to be completed, the types of fungi involved change over time. At an early stage of decomposition, while the bark is still present, many of the fungi are ascomycetes; as time passes, various basidiomycetes become more apparent. However, the basidiomycetes associated with an intermediate stage of decomposition are not the same as those associated with the later stages, when

FIGURE 12.5. Clump of fruiting bodies of chicken-of-the-woods, showing the mass of overlapping fruiting bodies that sometimes reach the size of a bushel basket.

the log begins to lose its original form (Figure 12.6). In fact, at this point in the decomposition process, the fruiting bodies of the basidiomycetes that are present (predominantly thelephores) are mostly thin and crust-like and occur out of sight on the lowermost surface of the log that is in contact with the ground. They can be observed only when a piece of the decaying wood of the log is turned over to expose the lower surface. At this stage, various mosses and lichens tend to be the most conspicuous organisms present on the upper surface of the log. Many of the ascomycetes found during an early stage of succession are members of a large and diverse group known as "pyrenomycetes." Many of these are dark brown to black in color and thus are not readily apparent unless one looks at some of the places on a log (such as the lower surface, near the ground) where they can be most apparent. Several examples of these inconspicuous fungi can be observed near the base of the fallen log in Figure 12.6. At the peak of fungal diversity, during the middle stages of wood decomposition, the rotting log can display a variety of interesting fungal forms that delight the eye of the hiker and provide food for familiar forest creatures such as box turtles and chipmunks (Figure 12.7).

As already noted, many plants establish mycorrhizal associations with fungi. Since the actual physical linkage between the plant and the fungus occurs underground, where the root system of the plant and the hyphae of the fungus are found, mycorrhizal associations are not necessarily apparent in nature unless one knows what to look for in a forest during late summer and early fall. The mycorrhizal structures located at root tips are just barely visible to the naked eye when roots are examined (Figure 12.8). There are two main types of mycorrhizal associations. In one, termed an "endomycorrhizal association" (Figure 12.9, b), the cells of the plant root are penetrated by the fungal hyphae (the prefix "endo" means that the fungus is largely inside the cells of the root). In the other type, termed an "ectomycorrhizal association" (Figure 12.9, a), the fungal hyphae invade the outer tissues of the root but don't actually penetrate the individual cells, and the hyphae form a covering ("sheath" or "mantle") around the root (the prefix "ecto" means that the fungus is largely outside the cells of the root). Endomycorrhizal associations are by far the more common of the two, but ectomycorrhizal associations are found in some of the more widespread and common trees of the Northern Hemisphere. Moreover, the kinds of fungi forming the two types of associations are different. Ectomycorrhizal fungi are predominantly basidiomycetes, while endomycorrhizal fungi belong to other groups. This is an important difference because basidiomycetes produce fruiting bodies that can be detected easily in nature, but this is not the case for endomycorrhizal fungi. In Arkansas, trees forming ectomycorrhizal associations include pine, beech, and oak, while trees forming endomycorrhizal associations include maple, cherry, and red cedar. In a forest in which only trees in the latter category are present, one would not expect to see the fruiting bodies of mycorrhizal fungi on the forest floor. That's because these tree species have endomycorrhizal roots, which don't involve basidiomycetes and thus don't produce fruiting bodies we would recognize. However, in forests with oak or pine present (which is often the case in the forests of Arkansas), fungal fruiting bodies

FIGURE 12.6. Decaying logs in three stages of decomposition, showing some of the fungi and other organisms typically present at each stage. A time scale for the progress of decay of the log is given by the growth of the oak seedling in the left foreground.

FIGURE 12.7. Decaying oak log at a typical peak of fungal diversity. Chipmunk and box turtle are among the animals that often consume fungal fruiting bodies.

such as mushrooms would be expected. For anyone who collects and studies fungi, consideration must be made of the trees present. The fruiting bodies of saprophytic fungi can occur in all types of forests as well as in non-forest situations.

The importance of fungi to forests and other ecosystems cannot be overstated. If fungi did not perform the essential ecosystem service of breaking down dead plant material (mostly dead leaves and woody debris) to release the nutrients required for life, the forest could not continue to exist. Moreover, most trees would have difficulty surviving in the absence of the mycorrhizal associations described above. However, as already noted, most of the fungi that are so vital to a forest ecosystem are not readily apparent, since they occur in the soil or within dead plant material and thus escape the attention of all but the most astute observers.

Many different kinds of insects and other animals feed on fungi,

FIGURE 12.8. Almost all forest tree roots have developed mycorrhizal associations with fungal species that appear as tiny elongated bulblets situated at root tips, as shown in this root sample from an Ozark white oak.

consuming either the fruiting body or, in the case of some microscopic organisms, portions of the mycelium. The fruiting bodies of fungi also offer shelter and a breeding place for some types of insects. For example, breaking apart an older fruiting body of one of the larger fleshy fungi will often reveal the presence of "wiggly white worms" that are actually the larvae of a group of small flies called "fungus gnats." Doing the same thing with the fruiting body of an old polypore usually produces a surprising number of different kinds of insects (often including beetles) and other small invertebrates. This would be the case for the older polypores on the decaying log shown in Figure 12.4. During periods when they are relatively abundant, the fruiting bodies of mushrooms represent an important item in the diet of some small mammals, particularly squirrels and chipmunks, and at times they are eaten by deer.

Almost invariably, one of the first questions asked about any wild fungus is whether it is edible. Although many fungi are edible, others are deadly poisonous. Unfortunately, there is no simple rule or test to distinguish edible from poisonous fungi, and unless a particular example has been identified with certainty (especially by an expert),

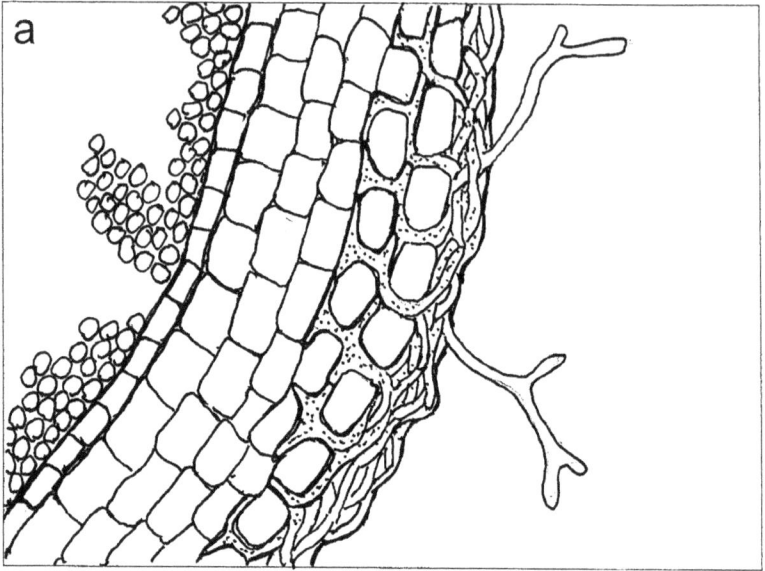

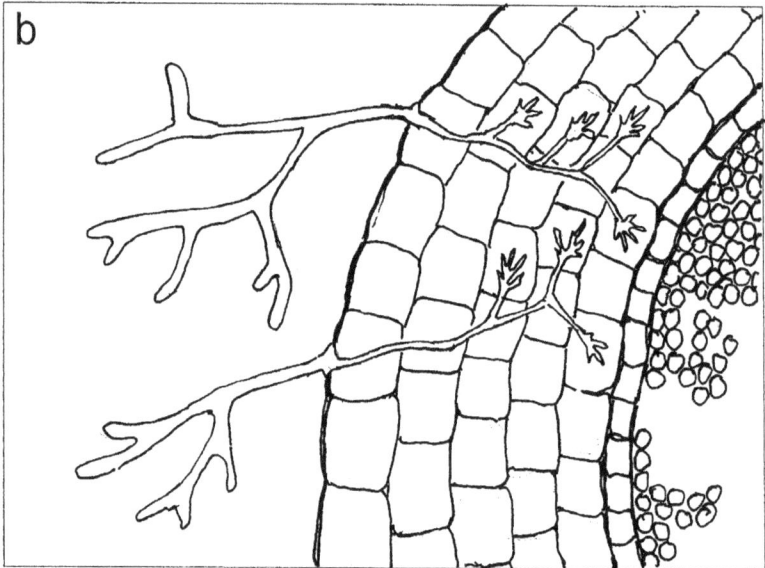

FIGURE 12.9. The two types of mycorrhizal association: (**a**) ectomycorrhizal and (**b**) endomycorrhizal.

it should not be regarded as safe to eat. Even those fungi known to be edible should be cooked before being consumed. Only a very limited number of fungi (the commercial mushroom commonly found as an item on salad bars is by far the best-known example) can be consumed raw (i.e., without being cooked first), and even these produce a gastrointestinal disorder in some people.

This is only a brief overview of the assemblage of fungi you might encounter on a walk in an Ozark forest—a limited introduction to the complex biological world found in the leaf litter and other plant detritus often ignored by hikers. The next time you hike down a mountain trail, pay attention to the leaf litter underfoot and the old logs and broken branches lying on the forest floor. The intriguing little mushrooms and intricately scalloped rows of small but often colorful brackets, projecting from standing dead trees and decaying logs and stumps, are literally the tip of a "fungal iceberg" largely hidden beneath the visible surface layer of the landscape.

Even after reviewing the many forms of woodland "mushrooms" and the vast world of fungi within that general category, there is yet another, very different, fungal life-form that provides a vast array of textures and colors for the hiker to contemplate. In addition to the forms discussed thus far, members of this other group of fungi are relatively common and often conspicuous in the forests of the Ozarks. These are the lichenized fungi, which make up most of what are referred to as "lichens." But what we recognize as a lichen is a "composite" organism in which a particular kind of fungus (almost always an ascomycete) is intimately associated with certain types of algae. The vegetative body (or "thallus") that results from the combination of these two entirely different organisms is a truly remarkable structure that bears little resemblance to either of its two component parts. Indeed, most lichens are so different from fungi that it would be hard to imagine the two being confused. In fact, the true nature of lichens as composite organisms was not comprehended until about a century and a half ago.

The lichen partnership formed by the two different organisms living together allows both to survive in conditions unsuitable for either partner alone. The algal part of the lichen produces organic molecules through photosynthesis that are utilized by the fungal part. In a simple sense, the alga is providing food to the fungus. In return, the fungus creates a favorable microenvironment for the algae that live within the thallus. The algae are protected during periods of desiccation, shielded from excessive solar radiation, and provided with mineral nutrients that are either extracted by the fungus from the substrate on which it occurs or deposited directly on the upper surface of the thallus from the atmosphere.

Their unusual partnership has allowed lichens to be widespread and enormously successful in some ecological situations that are not easily exploited by other organisms. For example, lichens are often the first macroscopic organisms to colonize bare soil and rock surfaces, and the term "nature's pioneers" is frequently used in this context. Geologists have developed a method known as "lichenometry," whereby the time over which a rock surface has been exposed is determined by calibrating the size of lichen colonies that have developed after that surface was first exposed to sunlight. Lichens also commonly occur on the bark surfaces of living trees, on old stumps and other pieces of decaying wood on the forest floor. These are the lichens that contribute to the appearance of virtually every tree and exposed rock we encounter on our outings.

It would be virtually impossible to find a place in the Ozarks absolutely devoid of lichens. However, unless you are attuned to the presence of these organisms, their contribution to the color and texture of the overall forest scene tends to be greatly underappreciated. Interestingly, since lichens are perennial in addition to being so common, they have been regarded as the ultimate "survival food" for humans stranded in the wilderness. For example, the famed British explorer Sir John Franklin and his party subsisted on boiled lichen, a menu item described as "rock tripe" in Franklin's journal, during a time when they were stranded in the barren wilderness of northwestern Canada. Although the appearance of such "fleshy" lichens is not particularly appetizing and their thallus is tough to digest, they are certainly easy to collect and at least nutritious enough to sustain life.

On your next hike, look at the bark of trees and the surfaces of boulders (Figure 12.10) to realize how much of the forest scenery is characterized by lichens. Almost every tree will have lichens present, even on the roughest bark plates of trees such as oak (Figure 12.10, a).

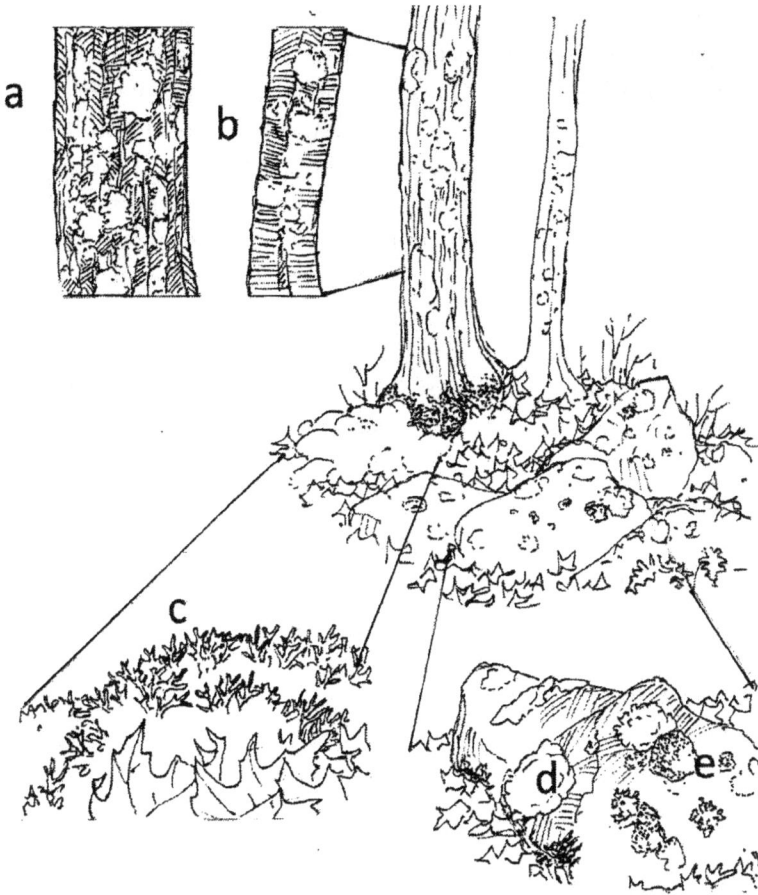

FIGURE 12.10. Lichens contribute to the textures and color of the forest scene along almost every foot of a typical forest trail, where relatively light-colored lichen patches mottle the rough bark of oaks (a) and the smooth bark of red maples (b), form soft cushions of "reindeer moss" on dry rock ledges (c), and appear as splotches of thick, light-colored lime green lichen (d) or thin dark gray lichen (e) on the weathered surface of exposed trailside boulders.

Close examination of smooth-barked trees such as relatively young maples, beeches, and Ozark chinquapins shows oval or round patches of contrasting color produced by the presence of lichen colonies (Figure 12.10, b). This patchwork of contrasting color makes bark that would otherwise be rather drab and uniform much more interesting to look at. An equally instructive exercise is to stop and take a close look at rock exposures produced by recent construction along roads. Compare the color and appearance of the freshly exposed rock with that of nearby boulders that have accumulated colonies of lichens that vary in color from light lime green (Figure 12.10, d) to almost black (Figure 12.10, e) to see how completely the lichens affect the appearance of rock surfaces. Patches of lichen also encroach on dry rock ledges as "reindeer moss"—hummocky, light gray-green cushions that can eventually carpet the edges of the exposed rock surface (Figure 12.10, c). This relatively light color and unique texture can form a dramatic contrast with the adjacent leaf litter and other lichen-mottled rock surfaces, offering an especially appealing photo opportunity for the attentive naturalist. Ozark forests would be much less interesting places to visit were it not for the many shapes and textures of lichen that decorate almost every rock, tree, and stump along the trail.

CHAPTER 13

Diseases and Pathogens

The trees and other plants of Ozark forests are subject to infection by parasites and pathogens—including fungi, insects, and even other plants—that produce damage and sometimes death. The major role of these various pathogens in determining the character of the woods is easily observed on our outings. As noted in Chapter 12, the distinction between "parasite" and "pathogen" is not absolute, and a particular organism may coexist with its host (and thus be considered a parasite) in some situations but kill its host (and thus be considered a pathogen) in others. Scientists surmise that forest trees invest a significant portion of their photosynthetic resources into the manufacture of chemical defenses to ward off the constant array of pathogens and herbivores to which they are exposed.

From an ecological point of view, the various fungi and/or insects that can cause the death of a tree are the most important pathogens, since they have the potential to affect the overall composition of an entire forest. In fact, the chestnut blight fungus, undoubtedly the best-known example as a result of its enormous economic impact on the forestry industry, essentially eliminated American chestnut from many areas of eastern North America during the first half of the twentieth century. In some of the areas that were most affected, American chestnut trees represented half or more of all stems present in the pre-blight forest and about half of the usable timber crop, according to industry estimates at the time.

Although the introduction of chestnut blight into eastern North America is thought of as having resulted in the tragic loss of a major, economically important tree in the Appalachian region, this disease also devastated the Ozark chinquapin, found in Arkansas, Missouri, and Oklahoma. The blight had spread through the southwestern

Appalachian region by 1940 and had reached the range of the Ozark chinquapin in the 1950s. The chinquapin (not to be confused with the chinquapin oak) was not an especially abundant tree, comprising only about 1 percent of the original forest cover as inferred from "witness tree" surveys (Chapter 5). However, the tree did provide a reliable, abundant, and nutritious nut crop for wildlife and humans, growing in groves on ridgetops and sandstone ledges (Chapter 10). The exact date of the blight's arrival can be determined by looking at the growth rings of oaks that are found next to the remains of a large chinquapin tree killed by the original blight pandemic. Data obtained in northwest Arkansas (Figure 13.1) show a release date of 1958 (the year when enhanced growth rate is indicated on tree cores) for oaks once competing with chinquapin, indicating that the competition was killed off in the preceding year (1957). Using the average growth rates of many trees to form a local growth chronology (see Chapter 7), we find no indication of especially favorable growth conditions (Figure 13.1). Therefore, the sudden increase in the growth rate of oaks adjacent to the remains of large, blight-killed chinquapin trees can be unambiguously attributed to the removal of competition when blight arrived in 1957.

The organism responsible for chestnut blight is a wind-disseminated fungus that kills the bark on the lower part of trees and saplings but doesn't kill the root system. Large chinquapins were able to resprout from the base but were less vigorous than smaller trees, and the weight of the dead trunk and the weakened root system usually caused the tree to topple over, pulling its roots out of the ground. Even so, chinquapin survives today in the form of what have always been small saplings or seedlings ("old seedlings") that continue to send up new shoots as soon as a stem is killed back by the blight. The situation is illustrated in Figure 13.2, which shows the appearance, in 2015, of once large chinquapin trees killed by blight in 1957, with a small surviving sprout in the background. The latter is growing at the base of a dead sapling that had been killed by blight a few years earlier, and this root system has probably been through several cycles of dieback and resprouting. Thus, the species is still an important component of the Ozark forest understory but has attained an especially difficult situation with respect to long-term survival. First, the

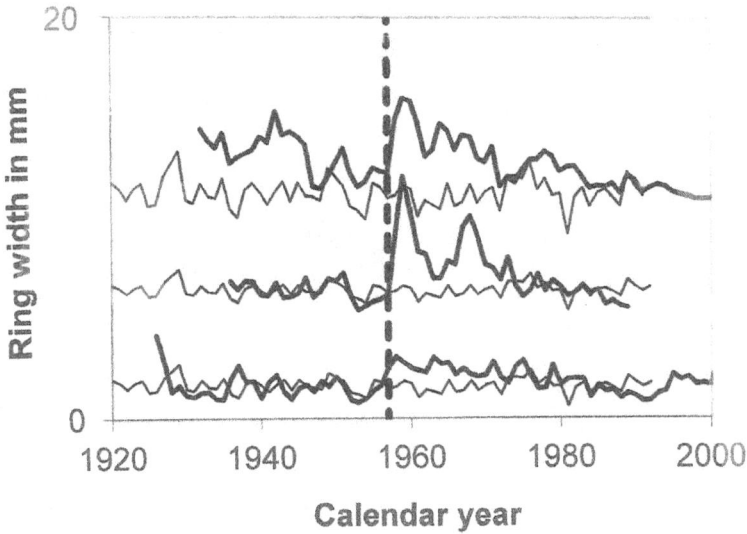

FIGURE 13.1. Ring-width measurements from three oak trees that grew for decades in competition with large Ozark chinquapins (on three different sites in northwest Arkansas) show how the elimination of that competition—starting in 1958 after the arrival of chestnut blight the previous year—affected growth. Individual ring-width chronologies (thick lines) are superimposed on the average chronology from many trees (thin lines) to verify that the sharp increases in growth rate starting in 1958 are not attributable to climate.

sprouts rarely become large enough to reproduce sexually as a way of evolving blight resistance. Second, the ability of most chinquapin root systems to withstand blight means that less blight-resistant individuals are not eliminated from the population. Fortunately, researchers at several universities are now working with the Ozark Chinquapin Foundation on a breeding program aimed at restoring the chinquapin to future Ozark forests.

In recent decades, one of the most economically important forest diseases in the Ozarks has been a blight often attributed to damage by the red oak borer (Figure 13.3). The red oak decline is a real puzzle, because the borer is a native insect and its most common victim, northern red oak, often manages to resist infestation. Similar red oak mortality episodes have been observed since 1950 in the eastern states and have been studied in great detail. The ultimate cause of

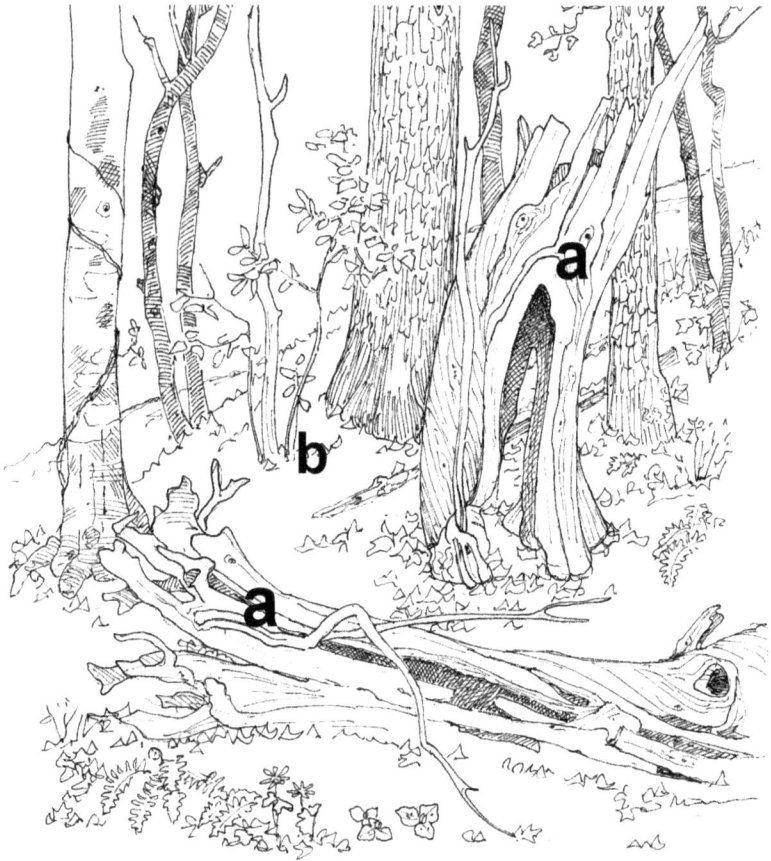

FIGURE 13.2. Typical upland forest scene in 2015, with the remains of what were once large Ozark chinquapin trees (**a**) and a living chinquapin sprout subject to periodic stem girdling by blight (**b**). Note that almost all the living sprouts are not adjacent to the large dead trees—–they are "old seedlings," established before blight removed the chinquapin seed source. Also note that the rot-resistant chemistry common to the wood of all species of chestnut trees has prevented moss and lichen from growing on the old chinquapin wood.

these episodes is thought to be natural stresses that weaken the trees and make them susceptible to the borer. For example, a red oak die-off in Connecticut has been attributed to trees being weakened by severe drought in the 1960s; and another, in Virginia after 1970, to defoliation by the introduced gypsy moth caterpillar. Unlike the examples of parasites and pathogens discussed elsewhere in this chapter, oak decline in the Ozarks appears to result not from a single factor but from several. First, upper slopes and ridgetops tend to be characterized by shallow, rocky soils that are not especially suitable for good tree growth. Second, many of the oaks that occur in these situations have probably reached an age of at least seventy to ninety years, when they gradually become somewhat less vigorous and thus more vulnerable to such stressors as short-term drought (as in an unusually long, dry summer), an extraordinarily cold winter and/or late-season frost, or the damage caused by repeated partial defoliation by naturally occurring leaf-feeding insects. Third, if an individual tree becomes stressed, this increases the likelihood that certain insects (especially wood-boring beetles like the two-lined chestnut borer or the red oak borer) or fungi (including such noteworthy examples as the *Armillaria* root rot fungus, a rather common root-decaying basidiomycete), already present in the forest community, can overcome that tree's natural defense system and invade or infect its trunk or root system. Thus, oak borer damage alone may not be the primary cause of red oak decline in the Ozarks, although the advance of the blight may continue once the population of the borers reaches elevated levels, so that even healthy trees are unable to resist the onslaught of that pathogen.

Dutch elm disease is also very destructive (Figure 13.4). It infects American elms (and, to a lesser extent, other species of elms), first killing individual branches and eventually resulting in the death of the tree (within one to several years). The fungus responsible for the disease is thought to have been native to Asia, but it was first studied seriously by scientists in the Netherlands, which accounts for the common name. Dutch elm disease, like chestnut blight, is caused by a member of the ascomycetes. However, the ecology of the two diseases is quite different. The spores produced by the chestnut blight fungus are spread both by wind and by any insect or bird that happens to come into contact with an infected tree. As such, the potential for dispersal is almost limitless, much to the detriment of chestnut trees. By contrast, the spores of the fungus responsible for Dutch elm disease are produced in sticky blobs. Moreover, while the actual spore-producing structures of the chestnut blight fungus occur on the outer surface of the bark of the infected tree, where they are exposed to the wind, those of the Dutch elm fungus occur beneath the bark and are

FIGURE 13.3. Ozark forest scene with the remains of northern red oaks (**a–g**) killed a decade earlier by red oak blight, which is commonly attributed to the red oak borer (**h**) but may represent a more complex disease process.

dependent on a particular type of insect (the elm bark beetle) to carry them away from an infected tree. Thus, Dutch elm disease involves two different types of organisms, one with the potential to kill the host organism and the other responsible for spreading it to new hosts.

Dutch elm disease was first reported from North America in 1930, and since then it has devastated native populations of elms. Although the total number of trees killed by the disease was lower than the

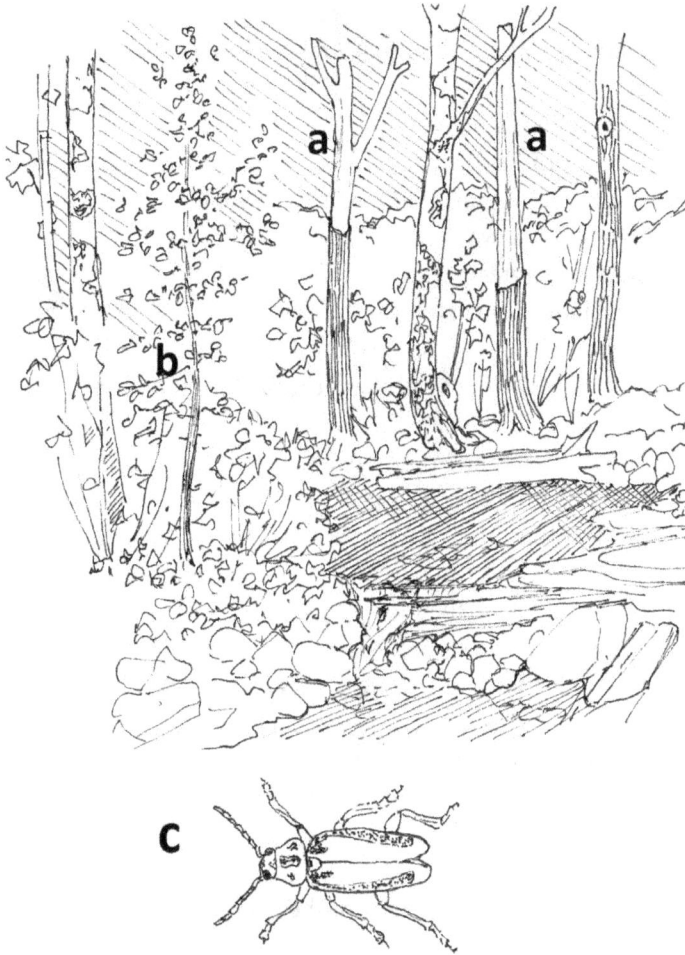

FIGURE 13.4. Blight-killed elms (**a**) in a typical Ozark streamside habitat where rapidly regenerating elm saplings resulting from an abundant seed source (**b**) ensure that elms will be part of the forest in the future; the elm beetle (**c**) is instrumental in transporting the fungus responsible for the disease from one elm to another.

number killed by chestnut blight, the effects of the disease were more noticeable. This is because American elm was a widely planted urban shade tree, and many streets in cities throughout the country were once lined with elms. Sadly, this is no longer the case. On the positive side, elm trees grow rapidly and produce seed readily, even now that elm blight is endemic, so that the process of natural selection is likely to develop resistant elm trees in the future. All three elm species native to the Ozark region are abundant today and likely to remain so into the future.

The emerald ash borer poses a severe threat to ash trees in the Ozarks (Figure 13.5). This bright green beetle, native to Asia and eastern Russia, is a recent introduction and was first discovered in North America in 2002. It was accidentally brought here in the ash wood used in shipping crates. In those areas where the beetle has been introduced, it is highly destructive, typically resulting in 100 percent mortality of ash trees. In this case, death results from the feeding activities of the larval stage of the beetle, and there is no fungus involved. The galleries (tunnels) created by the larvae feeding beneath the bark disrupt the flow of water and nutrients, effectively girdling the trunk and ultimately killing the tree. It takes several years for this process to kill a large ash tree, but unfortunately the results are inevitable. Because the ash borer is such a tiny insect, it is easily overlooked and difficult to collect in the forest. The most effective way to determine whether the borer is in an area before trees start to die off is to find open areas with sandy soils. A species of wasp inhabits burrows in such places and works hard to feed its larvae by catching insects and returning with them to the nest. An investigator can stand by one such set of burrows with a hand net and collect the insects that the wasps bring back. The emerald ash borer was reported in Arkansas for the first time during the summer of 2014, and it is still too early to know what impact it will have on the forests of the Ozarks, where white ash is a relatively common and widespread tree.

Another insect pest produces noticeable damage on trees that is never fatal but can disfigure specimen trees in parks and can introduce defects in otherwise straight-trunked timber trees (Figure 13.6). These pests are the thirteen-year and seventeen-year locusts (also known as "cicadas"), which spend most of their life underground and then erupt en masse at the specified interval to mate and lay eggs for the next generation. They tend to come in cycles and can damage the main stem (or "leader") of a young tree because the incision made to insert eggs often causes branch tips to die back. In young, regenerating stands of oaks, the locusts tend to pick the tallest and most well-formed trees for their egg deposits, selectively wounding what would

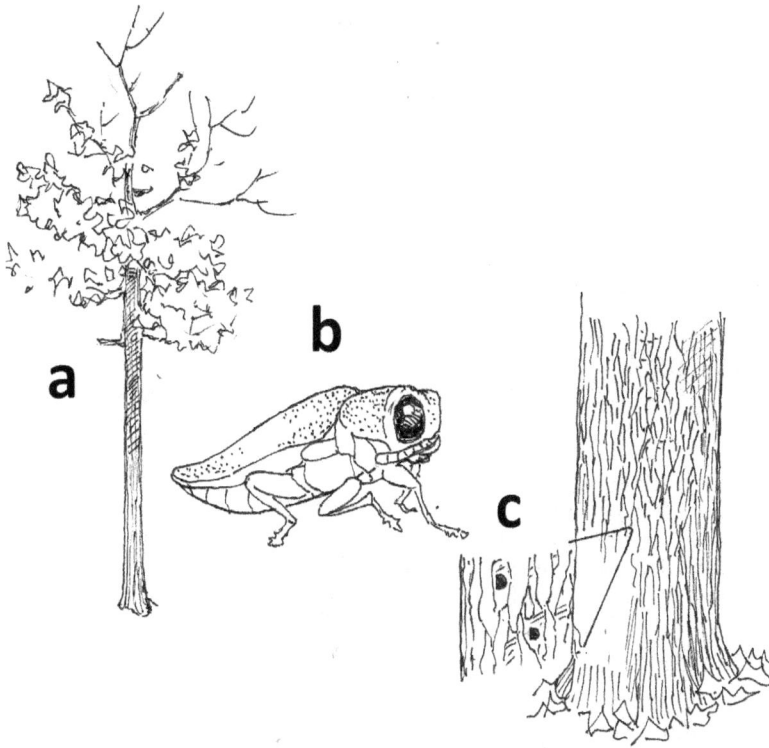

FIGURE 13.5. White ash (**a**) succumbing to an attack by emerald ash borers—the adult form of which (**b**) is iridescent green (indicated here by stippling)—in a woodlot in Connecticut, where the latter are actively eliminating ash trees from the forest. The tiny, eighth-inch-diameter, D-shaped holes (**c**) from which the borers emerge indicate the size of this diminutive pest.

otherwise be the best and most well-formed trees in the future forest. The brown and shriveled clumps of foliage on affected trees can be very noticeable in the forest, but most trees can generate a new leader and grow themselves out of the deformation.

"Heart rot" is a general term applied to numerous different fungi, many of which are polypores (discussed in Chapter 12), that cause decay in wood at the center of the trunk and larger branches of trees. The particular fungus usually enters a tree through a wound in its bark and becomes established in the heartwood (the portion of the wood located at the center of the trunk). As the heartwood becomes

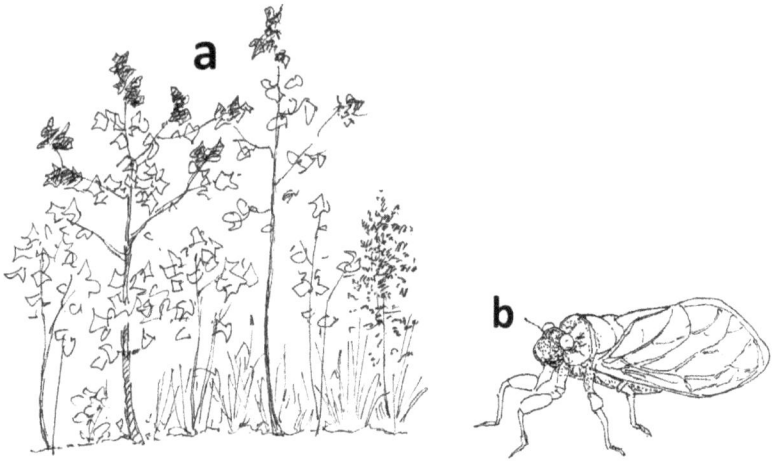

FIGURE 13.6. In early summer, the growing branch tips of the most vigorous young oak trees in an area of regenerating forest can be seen to have dark brown, withered leaves (**a**) where a recently emerged cohort of thirteen- or seventeen-year locusts (**b**) have inserted eggs into the soft wood, damaging the tender, newly formed tissue and incidentally deforming the shape of the most robust young trees in the area.

decayed, it first becomes softer and then may disappear completely, thus producing a hollow trunk. Hollow trees have always been an important resource for wildlife in the Ozark ecosystem. Bears, in particular, prefer large hollow trees for hibernation because such locations offer shelter away from the damp forest soil during the winter months in regions such as Arkansas, where relatively mild winters (compared to Minnesota or Maine) prevail. However, hollow trunks result in the tree becoming structurally weaker and prone to breakage when the trunk is subjected to high winds or some other stress. Beech and black gum are especially susceptible to this problem (Figure 13.7). Until this happens, the tree may appear perfectly normal, although the presence of the fruiting bodies of a fungus somewhere on the trunk is a pretty good indication that it may be affected by heart rot. Heart rot is a major factor in the economics of logging, since trees with this condition have relatively little value. It is also important in the natural growth dynamics of older forests, in which the collapse of mechanically weakened trees is an integral step in forest regeneration.

Fortunately, not all the pathogens and parasites associated with

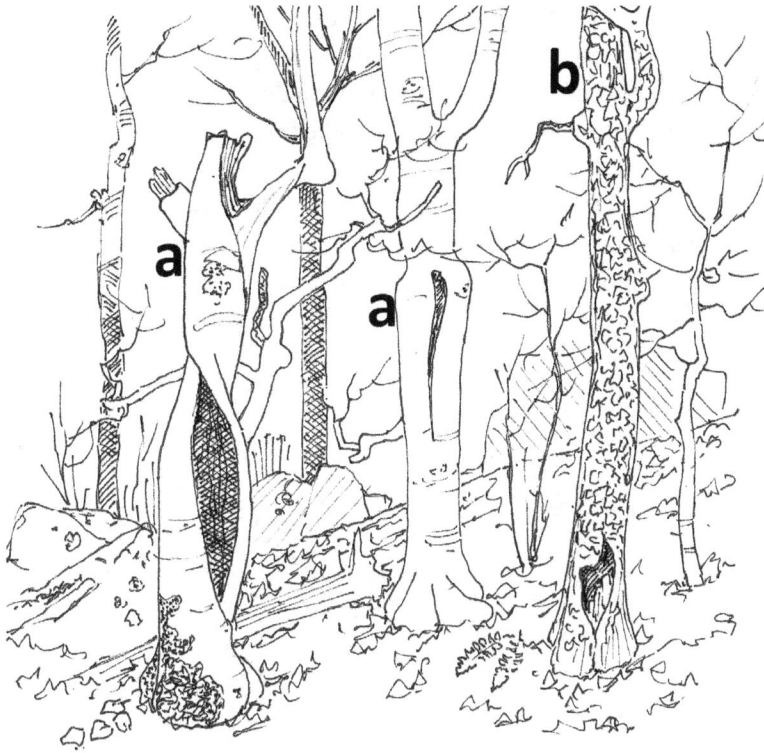

FIGURE 13.7. Large old-growth beech (**a**) and black gum (**b**) trees with extensive heart rot and breakage resulting from heart-rot-weakened trunks.

particular kinds of trees in our forests are not quite so deadly. In fact, some are relatively benign, at least most of the time. For example, red cedar trees are subject to infection by cedar-apple rust. The rusts are a group of fungi characterized by a life cycle that involves two different hosts. For cedar-apple rust, the two hosts are red cedar trees and apple trees (or sometimes crabapples in localities where apples don't occur). On red cedar, the fungus produces reddish-brown galls up to an inch in diameter on the branches of the tree (Figure 13.8). The surface of each gall is characterized by numerous small, circular depressions, each with a small, pimple-like structure in the center. In the spring, these structures produce gelatinous, horn-like, orange projections. The spore-bearing horns swell during rainy periods in April and May, and wind carries the microscopic spores to infect apple leaves, fruit, and young twigs within a radius of several miles of the infected tree. The numerous galls with their horn-like projections can make a heavily infected red cedar look somewhat like a decorated Christmas tree.

On an apple tree, the infections of the fungus occur on the leaves and fruit, and its characteristic, brightly colored spots make cedar-apple rust easy to identify. These begin as small, pale yellow spots on the upper surfaces of the leaves, usually appearing in late April or in May. These spots gradually enlarge, turning orange or red, and may display concentric rings of color. Later in the season, tube-like structures develop on the undersurface of the apple leaf, and these give rise

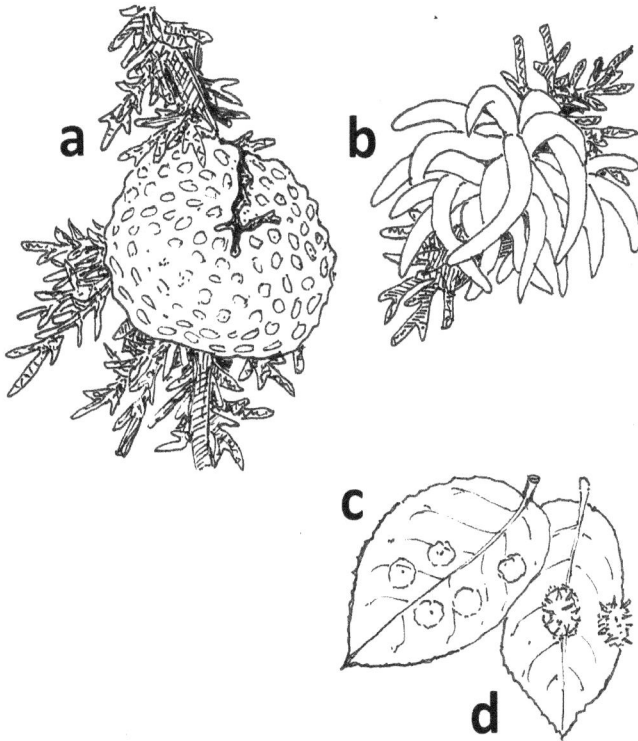

FIGURE 13.8. Galls of cedar-apple rust on a red cedar branch (**a**) where the orange telial arms arise from golf-ball-like depressions and swell to gelatinous, spore-producing arms (**b**) during wet weather; the spores settle on apple leaves to produce yellow spots (**c**), and spores are shed from the undersides of these leaves to reinfect cedar trees (**d**).

to spores that the wind can carry back to red cedars, thus completing the infectious cycle. Because both types of trees have to be present in the same general locality for this to happen, one measure that has been taken to control the spread of the disease in areas where apples are an important crop is to remove all red cedar trees. Unfortunately, red cedar is so ubiquitous in the Ozarks that this is not possible, and a once vibrant apple industry has been largely abandoned.

Black knot, a common fungal disease of black cherry trees throughout the Ozarks, causes rough black swellings (or "knots") to develop on the branches of infected trees. These are easily noticed, especially during the winter months, after the trees have lost their leaves. In fact, it is possible to identify a black cherry tree from some distance away by the conspicuous black swellings. Although black knot disease may not kill an infected tree, the latter will tend to lose vigor, become increasingly unproductive in its flowering and production of fruit, and become more susceptible to injury from the environmental stresses of winter. The disease is occasionally found on other, closely related trees (including commercial cherries, peaches, and apricots) but seldom presents a serious problem. Because black cherry is a relatively uncommon tree in Ozark uplands, and because these knots are most evident on younger trees in abandoned fields, this disease is not often encountered on a typical hike.

A much more common gall affects hickory and sometimes oak trees (Figure 13.9). The fungus involved is a member of the genus *Phomopsis* (a type of ascomycete), and the condition it produces is referred to as "Phomopsis gall." The galls appear as round swellings on the branches or more rarely the trunk of an infected tree and can occur singly or in clusters, with a single gall often reaching several inches in diameter. When multiple galls occur on a single branch, the latter may be girdled and killed. The galls die after several years and turn black.

During the winter months, it is impossible to overlook the mistletoe that forms dense masses of bright green leaves, commonly occurring on otherwise bare branches of trees in Ozark forests (Figure 13.10). Mistletoe lives on a wide range of host trees, including maple, oak, hickory, black walnut, beech, box elder, ash, dogwood, persimmon, Osage-orange, sycamore, black cherry, and black locust. Technically, mistletoes are "hemiparasites," because they are capable of carrying

FIGURE 13.9. Mature black oak infested with *Phomopsis* galls—showing almost every branch full of galls, yet with no other obvious effect on the healthy-looking tree—and detail of a gall-covered branch (with oak leaf for scale).

out photosynthesis on their own in addition to obtaining water and some nutrients at the expense of the host tree. Mistletoe plants "tap into" the tissues of the host plant by means of specialized roots called "haustoria" (singular: haustorium). The presence of mistletoe on a branch results in a reduction in the growth of that portion of the tree, and heavily infected trees may be killed in exceptional instances.

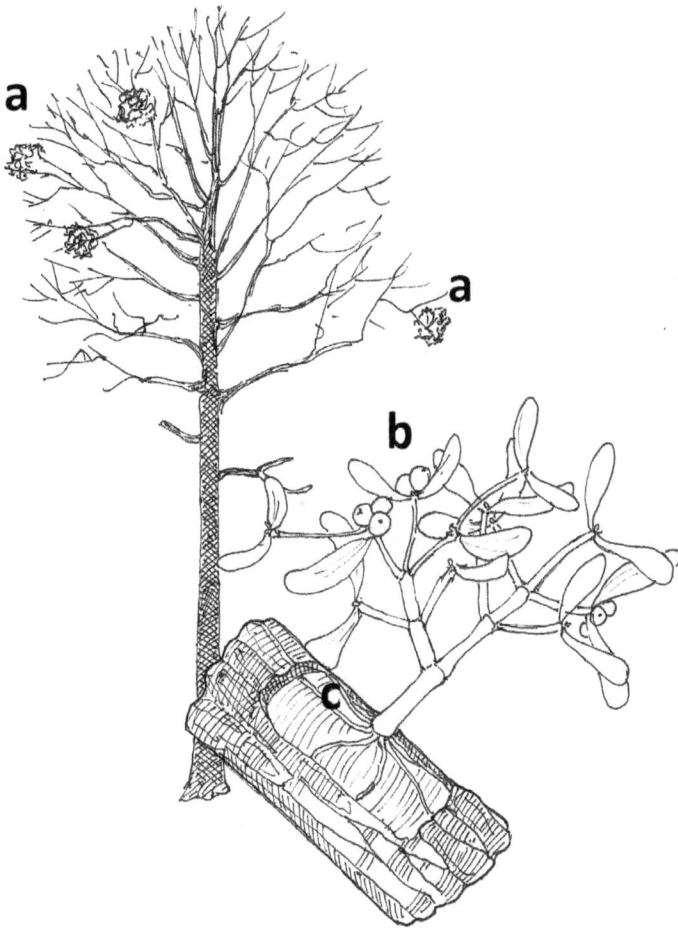

FIGURE 13.10. Mistletoe is a parasitic plant that grows by tapping into the vascular system of tree branches and appears as spherical masses of foliage in the crowns of trees such as the southern red oak (**a**). Sticky seeds from the mistletoe's berries (**b**) are transported by birds and lodge on branches where mistletoe roots can grow into the cambium of a tree branch (**c**).

The leaves of mistletoe are simple, opposite, and evergreen. Both the stem and the leaves have a thick, waxy covering ("cuticle"). When a portion of the plant is collected, it remains fresh-looking for several days and thus is ideal for use as a Christmas decoration. The tradition of hanging mistletoe in a home goes back to ancient times, when the plant was thought to possess mystical powers that brought good luck and warded off evil spirits. Mistletoe was used as a sign of love and friendship in Norse mythology, the origin of the custom of kissing under mistletoe.

In the fall, it is not unusual to find what are known as "oak leaf galls" among the dead leaves on the forest floor. These round structures are brown and range from an inch to slightly less than two inches in diameter (Figure 13.11). Oak leaf galls are sometimes mistaken for puffballs (see Chapter 12) until they are picked up and examined more closely, when it can be noticed that they weigh very little; have a thin, papery outer covering; and are filled with a diffuse, spongy mass of leaf tissue—all of which are features not associated with puffballs. Although initially green while still developing (and before they fall to the forest floor), oak galls are formed as a result of chemicals injected into leaf tissue by the larvae of certain kinds of gall wasps. An adult female wasp deposits a single egg into a developing leaf bud on an oak tree, and the larva that hatches from the egg then feeds on the gall tissue that results from the abnormal growth response of the leaf. Ultimately, the larva gives rise to an adult wasp, which emerges through a small hole in the wall of the gall. Although relatively large and somewhat spectacular in appearance, oak galls seem to cause no measurable harm to an oak tree. Gall-like growths also can occur on the trunk and branches of oaks and a number of other types of trees, and their presence can be quite detrimental. Some of these, like the oak gall, are caused by tiny wasps, but others are the result of infection by a fungus or bacterium.

Another type of abnormal growth that can be observed in our forests is known as a "witch's broom" (Figure 13.12). This is the name applied to a dense mass of small branches that arises from a single point in the canopy of a tree, the resulting structure having the general appearance of a broom (or some type of bizarre bird's nest). Witch's broom is not caused by a specific type of organism but may arise when the developing branch of a tree becomes infected by certain fungi, insects, non-insect arthropods, or plant viruses, the most common of these causes varying from one tree genus to another. The abnormal growth produced is not a temporary phenomenon but is likely to persist for the life of the host tree. A similar abnormal process is responsible for the massive burls that sometimes appear on the trunks of oaks and other trees (Figure 13.13). Folk artists in the Ozarks some-

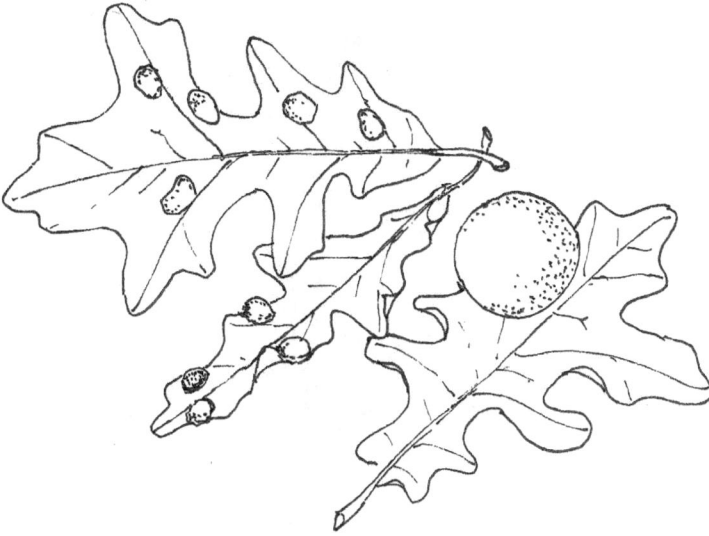

FIGURE 13.11. Typical oak leaf galls vary from small reddish pods to golf-ball-size structures.

times specialize in carving intricately textured bowls from burls that are cut from burl-bearing trees.

During the summer months, it is not unusual to encounter a living plant in which some of the leaves are characterized by the presence of a regular—but usually somewhat meandering or serpentine—pattern of light-colored markings that stand out in contrast to the green background of the leaf (Figure 13.14, a). This distinctive pattern is produced by a member of a group of insects known as "leaf miners." A leaf miner itself is the larval form of an insect and lives within and feeds on the tissue of the leaf, creating what is actually a feeding tunnel while doing so. The vast majority of leaf miners are moths, sawflies, and true flies, although a few are beetles and wasps. Since the larva lives within the leaf, it is afforded some degree of protection from predators while having easy access to a convenient food supply. The precise pattern formed by the feeding tunnel is often distinctive enough for an expert to identify the insect involved without having seen it. The larva leaves its "frass" (droppings) behind in the tunnel as it moves throughout the leaf. Although conspicuous, the damage

FIGURE 13.12. "Witch's broom" appears as a dense ball of needles and fine branches attached to a large branch in the crown of a shortleaf pine. Note the contrast between the compact mass of needles in the broom and the normal foliage on other branches.

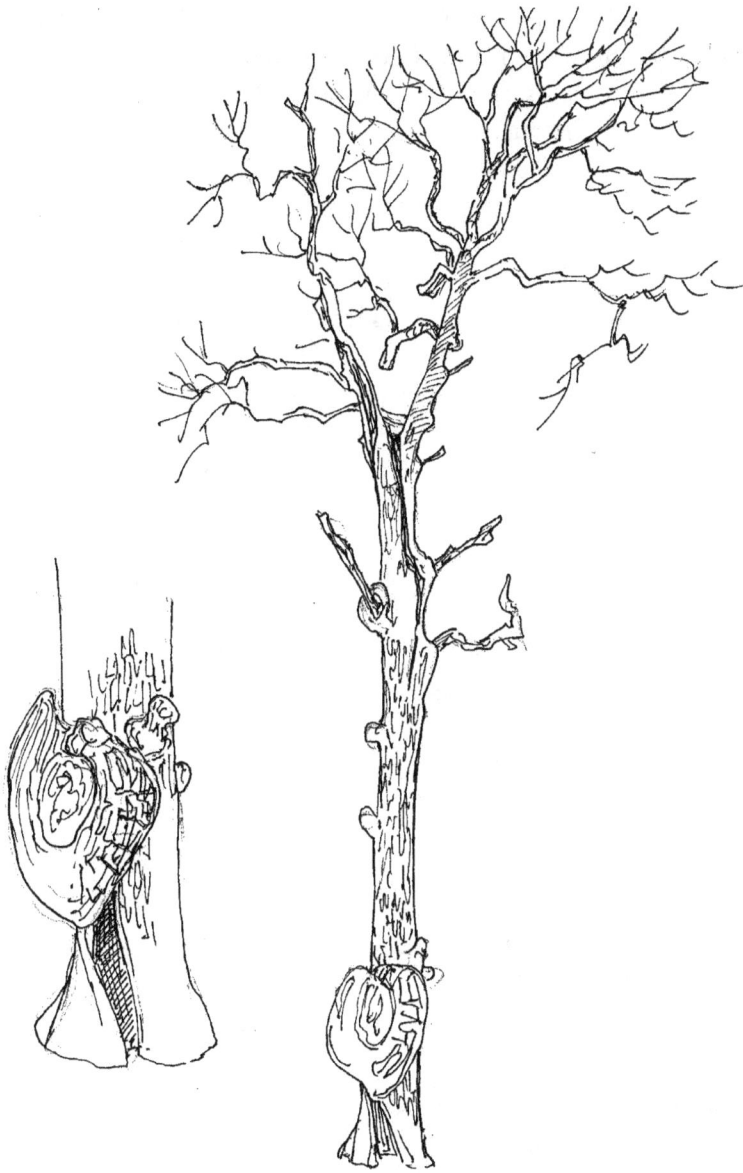

FIGURE 13.13. Burl attached to the trunk of a large white oak; such large, knobby growths are attributed to the same process of disrupted growth that produces a "witch's broom."

caused by a leaf miner is relatively minor and has little overall effect on the health of the plant. However, there are many other insects that consume leaves, and occasionally a major defoliation event will have serious consequences for the trees involved (Figure 13.14, b, c). Scientists have noted that trees can use airborne chemical signals to let surrounding trees know that they are under attack and to prepare themselves accordingly. However, the inevitability of autumn leaf fall means that our deciduous trees do not expend nearly the amount of resources in leaf defense that are invested by tropical trees, on which leaves typically remain in service for more than a full year.

Unfortunately, at some point in the future, the pathogenic, fungus-like organism *Phytophthora ramorum* could pose an even

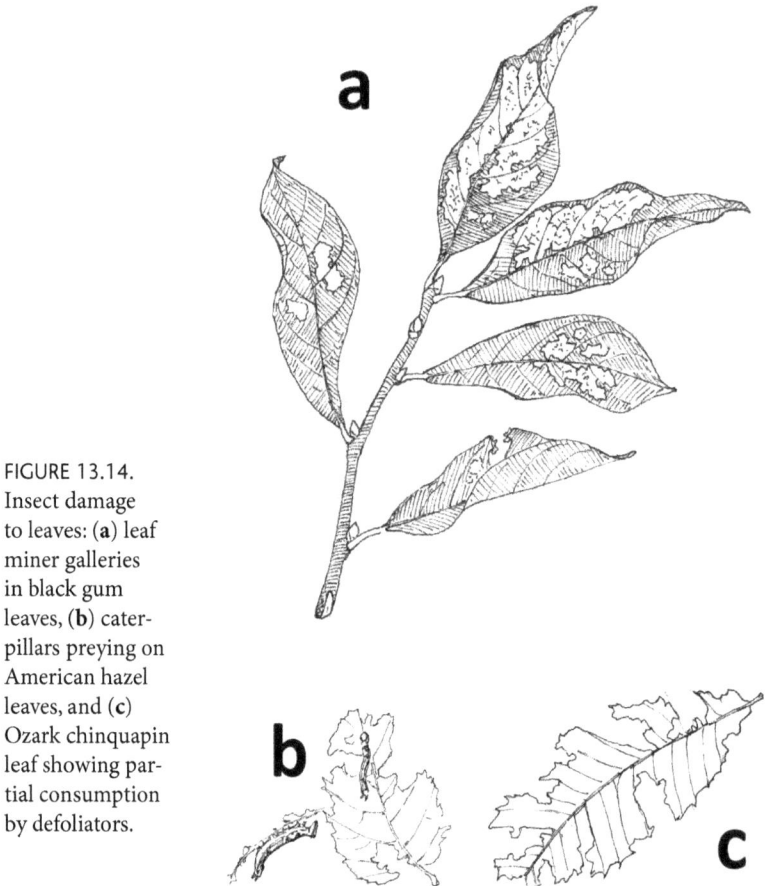

FIGURE 13.14. Insect damage to leaves: (**a**) leaf miner galleries in black gum leaves, (**b**) caterpillars preying on American hazel leaves, and (**c**) Ozark chinquapin leaf showing partial consumption by defoliators.

greater threat to oak trees in the Ozarks than chestnut blight and red oak decline. This organism, which results in a disease known as "sudden oak death" (now of great concern in California), infects oaks and certain other members of the oak family as well as some other plants. It is responsible for killing more than a million trees in the forests of California and is also known from Europe. Once infected, a tree usually dies rather quickly. Because it doesn't produce wind-dispersed spores, *P. ramorum* doesn't spread nearly as readily as diseases like chestnut blight. Indeed, it tends to depend largely on humans to reach a new locality. However, if the disease were to reach the Ozarks, there is little question that it would be devastating. Even if sudden oak death itself doesn't arrive in our area, other *Phytophthora* species (collectively known as "water molds" or "ink disease") are already here, and the Forest Service currently rates these as a major threat to the health of Ozark forests.

Another pathogen with potentially serious results for the southern and eastern Ozarks is beech scale disease, which has devastated forests from New Brunswick, where it was introduced more than a century ago, to Illinois, where beech trees are now being attacked. The disease is caused by the introduced *Nectaria* scale insect, which sucks sap from lesions in beech bark. This is not a serious detriment to the tree on its own, but the lesions allow a native fungus to gain entrance, which does have a serious effect on the tree. The damaged bark, cracked and unsightly, allows various insect pests to gain entry (Figure 13.15). One of the most important of these invaders is the wood-digesting carpenter ant. Not only do the ants weaken the wood, but they also attract pileated woodpeckers, which characteristically tear large holes in the trunks of the trees. The result is a snapped-off snag such as that in Figure 13.15. Then, because beech can sprout from its roots prolifically, a dense stand of beech saplings forms, completely changing the character of the forest environment. Since the scale spreads so slowly, it will be a decade or more before this pathogen arrives at the eastern edge of the Ozarks. Beech is a component of Ozark forests only in the southern and eastern parts of the region, but the tree is a distinctive part of certain habitats that will be severely altered when the *Nectaria* scale insect eventually crosses the Mississippi.

Careful examination of the leaves of living trees and shrubs in

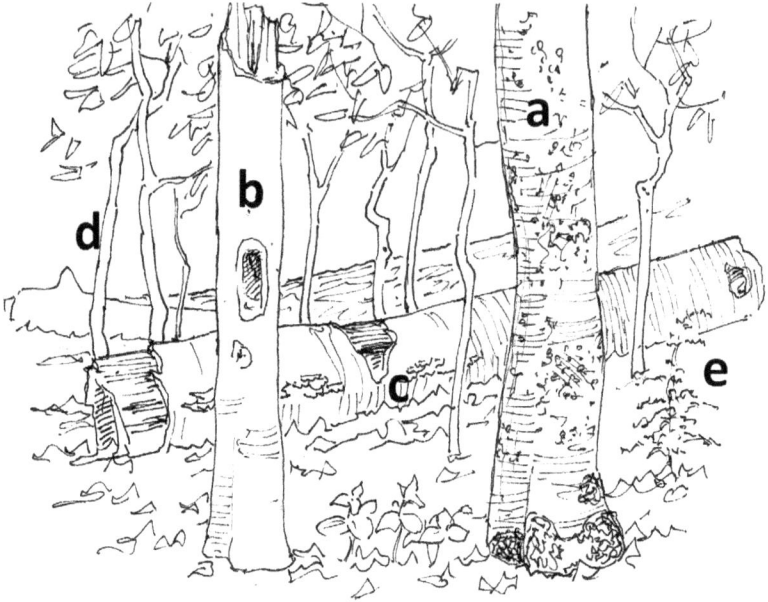

FIGURE 13.15. Ohio beech stand infested with beech scale disease, showing injury and damage to trunks of large trees where fungal blisters deform the bark (**a**) and allow infestation of weakened trees by insects that attract pileated woodpeckers, which produce large rectangular holes (**b**). The stand becomes littered with the fallen trunks of large trees (**c**) while stressed beech root systems generate many root sprouts (**d**). The small coniferous seedling on the right (**e**) is a hemlock, which is also now subject to a devastating introduced disease but does not inhabit the Ozarks.

Ozark forests, especially during late summer, will almost invariably reveal one or more "leaf spots" that stand out in marked contrast to the rest of the leaf (Figure 13.16). These spots can vary considerably in size and color, depending on the type of plant involved, the specific organism producing the spot, and the stage of development of the organism. Leaf spots are most often some shade of brown, but some are tan or black. Concentric rings or dark margins are often present. The vast majority of these leaf spots are caused by various fungi, but some are caused by insects or bacteria. Over time, individual spots may combine or enlarge to form blotches, and severely inflected leaves usually die. However, leaf spots appear to have no major detrimental effect on the plant as a whole. One of the more common and distinctive leaf spot diseases is the "tar spot disease" of maple. The name of the disease is derived from the shiny, black, raised areas, resembling splattered tar, that appear on the upper surface of a maple leaf. Like most other leaf spots, this disease seems to have little effect on the tree itself.

The vast array of pathogens and predators that assail Ozark forest trees today are the latest manifestations of an evolutionary battle that has been going on ever since plants first invaded dry land more than four hundred million years ago. Many of the trees in our region are descendants of temperate deciduous trees that extended over the entire Northern Hemisphere as part of the Arcto-Tertiary Flora some fifty-five million years ago (Chapter 10). That ancient forest would have contained trees such as oak, beech, and maple that we could easily recognize as being related to trees we see on our excursions in the Ozarks. The interactions between tree hosts and their pathogens that we observe today are the latest phase of a form of biological warfare that has been waged over deep geologic time. Some biologists speculate that only a small fraction of the solar energy that a tree captures

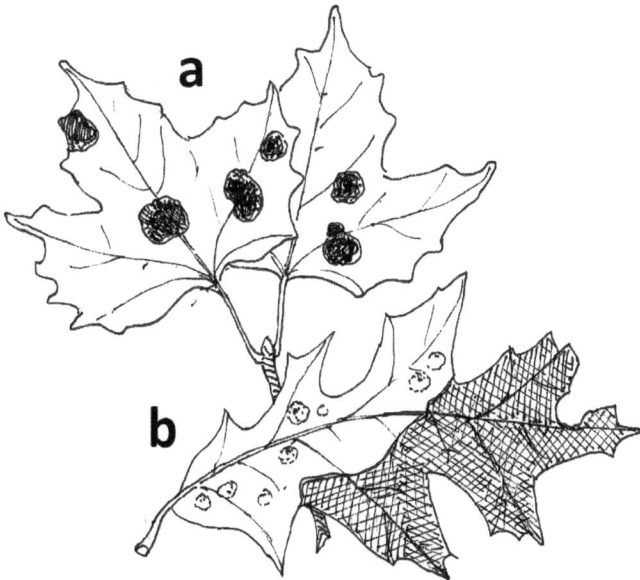

FIGURE 13.16. Leaf spots (**a**) on sugar maple leaves ("tar spot") and (**b**) on black oak leaves.

by photosynthesis is actually used in growth. They suggest that a much greater portion of a tree's resources are expended in creating the chemicals and tissue structure required to fend off leaf predators, wood borers, and other pathogens. It has even been speculated that the brilliant fall foliage we enjoy every autumn is a way for a tree to announce its blush of health and thus warn off potential enemies. At the same time, some insect pests have learned how to recognize chemical signals given off by trees in distress, using that cue to home in on victims with reduced defenses. Such complex chemical interactions are suspected to be part of the relationship between red oak trees and the red oak borer. So, as peaceful and relaxing as a stroll in the forest can be for the hiker, there is a furious chemical and mechanical battle going on all around us whenever we venture into the Ozark outdoors.

CHAPTER 14

Conservation Issues

We conclude with some thoughts about conservation and the preservation of our Ozark landscape. The preceding chapters are filled with the kinds of insights needed in forming well-thought-out conservation objectives, and the science of forest ecology continues to build our understanding of the forces that shape the landscape. Any discussion about policies and political actions to preserve forests should be based on this body of knowledge and inquiry. A good starting point in such a discussion is to ask exactly what it is we seek to preserve. Many of us think of conservation as the preservation, or even restoration, of a kind of "forest primeval." There is a problem with this, in that it remains unclear exactly what kind of forest that actually was (Chapter 5). But the ideal image of an unspoiled wilderness has always been a moving target. We once thought of the North American climate as being relatively static but punctuated by four major glacial episodes, triggered by some undetermined cause. We now know that orbital oscillations have been constantly modulating climate in an endless series of cycles extending as far back in time as can be observed in the fossil record (the "Milankovitch cycles"; Chapter 2). Moreover, humans had been interacting with the landscape for at least ten thousand years before European settlement of the Ozarks. The first humans arrived in North America more than twelve thousand years ago, and there is good evidence that their descendants made profound ecological changes to the ecosystem long before Europeans first set foot on the continent. As such, the first step in establishing a conservation ethic is to understand that there is no simple, static ideal of the unspoiled natural landscape. Our landscape has always been the product of a dynamic equilibrium, modulated over time as the annual cycle of the seasons varied over Milankovitch cycles and the prevailing disturbance regime drifted along its course (Chapter 5).

Our current natural landscapes are likewise subject to a dynamic process in which various environmental forces and ecological tolerances interact over various time scales. This means that the establishment of wilderness—as embodied in National Forest Wilderness Areas—cannot mean simply returning the landscape to its presettlement condition. Too many of the forces acting on the landscape have been permanently changed from even what they were a few centuries ago, so that kind of pristine condition cannot possibly be attained. "Wilderness management" is a nice conceptual tool for the conservation movement, but it really means managing substantial areas of land so that people can enjoy being immersed in an environment in which the "hand of man" is as little evident as possible. The spiritual value of such experiences, though separate from the science of forest ecology, is an important aspect of our outdoor culture. As such, we can accept wilderness management as a worthwhile goal for enhancing our enjoyment of the natural world. But even here the devil is in the details. The minimally intrusive nature of wilderness management includes a lack of trail maintenance and of interior access roads. This prevents the wear and tear of heavy use that other, more accessible areas receive, very often keeping the wilderness area relatively pristine and as close to wild as possible, given prevailing conditions. However, protection is not guaranteed if there is a specific feature, such as a spectacular waterfall, that attracts many visitors and is located not too far off the beaten path. When simply managed as wilderness, the area will still see a lot of foot traffic, and the braided network of informal trails and heavily used picnic spots around the waterfall will significantly degrade the area. In that situation, a well-designed access trail leading to a hardened viewing platform would obviously be preferable.

When considering how best to preserve our Ozark mountain environment, we must consider the forces that act on our landscape today and how we might manage them. Some of the factors that prevailed before settlement are still acting today: alternate periods of drought and heavy rainfall, occasional severe windstorms and tornadoes, freeze–thaw cycles during the winter months, epidemics of native forest diseases such as heart rot, or oak decline associated with borers. But other forces have been added or greatly changed: invasion by alien species, introduction of new tree diseases, increased erosion from construction and vehicle use, fragmentation of forest parcels by fields and suburban areas, elimination of large predatory animals, a continuous rain of non-point-source chemicals from upwind industrial sources, and a highly altered fire regime. Human beings are now an important part of the Ozark environment and have to be accommodated in our conservation policies (a pristine landscape is hardly worthwhile if we cannot be there to enjoy it). Conservation, then, involves managing the forces that act on our landscape, in order to accommodate our needs and provide for the enjoyment of the natural world around us while preventing ecological degradation that could threaten our very existence.

Given these considerations, what are the major conservation issues in the Ozarks? The state of our basic forest association, the oak–hickory forest (Chapter 3), has to be a major focus of our concern. The forests we have today consist of blocks of national and state forests that are large enough to function as ecological units, and they are clearly changing from an oak-dominated community as oaks fail to regenerate (Chapter 5). Mesic species such as maple and beech are expanding their role and the general consensus is that fire exclusion has been a major factor if not the primary cause of these changes. Restoration of fire in the form of controlled burns is a common practice today, but it may be that fire exclusion has gone on for too long to be easily reversed. Mesic trees and cedars are easily injured by fire when they are young and their bark is very thin, but they are much more difficult to eliminate when trunks get to be several inches thick and root systems have become so deeply entrenched that they can survive to resprout. Even if only a few of these mesic trees survive the burn, they serve as outlying seed sources for resuming the infiltration of the upland forest between periodic burns. A few expensive "brute force" treatments of the landscape may be needed to reverse the spread of mesic tree species and red cedar. This is being done, for example, at the Pea Ridge National Military Park in northwest Arkansas, where there is a perceived need to restore the landscape to an approximation of what was present at the time of the battle in 1862. This requires manual cutting and removal of large cedar trees over many acres of abandoned fields and disturbed forest to produce a condition that can be realistically maintained by a future program of regular controlled burns. This expensive and disruptive process is warranted for only a few, specially selected locations. Continued experimentation with thinning and burning at various scales may yield new ways to guide existing forests into better oak regeneration. It is also important to recognize that many recreational needs can be met by any healthy forest, and some areas can simply be allowed to evolve into mesic woodlands, with hardwoods eventually overtopping all but the densest groves of cedar.

Another important conservation concern is the loss of glades and other relatively small, open areas that serve as habitat for rare or endemic species, which are either outlying populations of prairie inhabitants or locally adapted to this one isolated type of environment. The long period of fire exclusion has made it especially hard to restore glades. Manual clearing of trees from the margins of glades may be required to restore them to such a condition that regular controlled burns can maintain the community. All available research confirms that glade plant communities (and the animals that depend on them) cannot persist without periodic burning of the organic soils that build up around the edges of rock exposures. Restoration and relocation of species to restored habitats may be needed so that an increase in the numbers of island habitats can ensure the viability of certain critical species. Corridors where small bits of open glade allow connections to exist between larger areas may also help preserve the gene base of some species.

Secluded ravines with remnant old-growth forest, another some-what limited resource, are highly valued by hikers and other outdoor enthusiasts. They have survived in this condition because of their inaccessibility, yet this very ruggedness—and the fragility of moss-covered ledges and boulders—makes such areas especially susceptible to damage by informal trails developed for human access. Muddy hillsides are difficult to traverse, and wide swaths of disturbed soil can be left when groups attempt to enter these areas. Some of the most popular of these secluded ravines need to be monitored. If the scenic rock gardens (Chapter 4) that visitors come to see are being degraded by too much traffic, conservation will require management. In the case of the most popular sites, it would be best to construct hardened, erosion-resistant trails and viewing locations to provide access while limiting damage to the most spectacular places.

The presence of invasive species is one of the most important changes affecting the dynamic forces in our modern Ozark woodlands. Invasive vines and shrubs such as Japanese privet, Japanese honeysuckle, and Asian bittersweet have completely altered the appearance of woodlands in suburban areas (Chapter 4; Figure 4.9). Feral hogs have likewise changed our landscape over a wide area, and not just in the vicinity of roads and communities. These invasive species and several others have become so deeply embedded in our landscape that they cannot be removed except locally and with significant investments of time and effort. We should therefore carefully choose which battles to fight. With a little public education and outreach, groups of volunteers might be assembled to grub out privet and honeysuckle bushes from local parks just to preserve the natural appearance of a particular area. Vigilant patrolling of the fringes of more pristine areas should probably be given priority. At the same time, we recognize that species like honeysuckle and privet provide a significant food source for animals and birds, and we can just accept their presence in some of our woodlands. Care should be taken that other possibly invasive species are not introduced to compound the problem.

One particular case of an invasive species' proliferation remains out of sight because it occurs entirely underground. Earthworms were not originally part of the natural organic recycling regime on this continent—they were incidentally introduced by soil clinging to plant and tree specimens brought here from Europe. Before earthworms were present in North America, leaf litter decay and organic soil processes were carried out by fungal activity alone. When earthworms continuously churn and process soil organic material, the activity of soil bacteria becomes much more important in the nutrient cycling in the subsurface. Thus, a major revolution in soil nutrient cycling is taking place beneath our feet, and the consequences of that revolution are just now being assimilated.

Many of us who enjoy the outdoors think of pollution as the one main threat to our forested waterways. That threat is largely out of the domain of forest management, except in the vicinity of campgrounds, picnic areas, and forest-access parking lots. However, there are other important issues related to keeping streams and rivers in the condition we desire. In Chapter 8, we considered the dynamic aspects of streams and small rivers—for example, a mountain stream interacts with the sediments it transports. Streams modulate the downstream movement of products of erosion from mountain slopes. Most visitors to our Ozark forests probably think of streams as a kind of permanent geographic feature, but the channel we see on a given day represents a dynamic equilibrium, subject to change that can sometimes be quite rapid. Small changes—such as those imposed by informal access points or vehicle crossings—can make a large difference, driving the stream channel out of its natural equilibrium. A local but major change in a stream channel's structure or slope can quickly propagate upstream and downstream over relatively large sections of the waterway. While many users of Ozark woodlands assume that scenic stream channels are fixed natural features, special attention needs to be paid to the design of streamside campgrounds, access points for floaters, and trails around waterfall-viewing areas. It is a simple fact that a stream channel's structure can be in a delicate balance, easily upset by relatively small changes in the drainage basin.

Another important consideration in conserving the landscape is the recognition of old-growth forest, some of which may be hiding in plain sight because a specific site may not be capable of producing trees suitable for logging or is otherwise not suitable for development. These sites can be recognized by the character of the trees growing there (Chapter 6). Sites that have been generally free of disturbance can also be recognized by the presence of certain herbs and wildflowers that require seclusion and have great difficulty spreading into new locations. Examples include relatively uncommon species such as ginseng and ladyslipper orchids, but also more familiar flowers such as bloodroot, hepatica, and dutchman's breeches (Chapter 11). Some of these sites may be located in and around suburban areas where they would be especially useful for recreation trails and could be preserved as a valuable community resource. Indeed, with such little pristine land available in the vicinity of developed communities, these last fragments of effectively wild landscape should be actively sought out and identified for preservation.

This book is intended to give hikers and other outdoor enthusi- asts a better appreciation of the Ozark landscapes they already greatly enjoy. Many of us belong to conservation organizations such as the Ozark Society or the Sierra Club, which advocate and work for pres- ervation and conservation. But we do this in a time of limited financial resources, and conflicting interests are involved in just about every conservation issue. So it makes sense to think carefully about the truly important objectives in preservation. We hope this short review has helped identify the issues that really matter, so that the most can be accomplished with the resources currently available.

LIST OF SPECIES
MENTIONED IN THE TEXT

It should be noted that there are two types of names used for organisms, the "common name" and the Latin (or "scientific") name. Many common names are used rather generally, not necessarily in reference to a specific type of organism. Moreover, many widespread organisms have multiple common names that vary between regions and languages, which can lead to confusion. Scientists use standardized Latin names—the Linnaean system of classification—to avoid such confusion. This system has levels that are useful in understanding the evolutionary relationship of one organism to another and to other groups of organisms. These levels of classification are species, genus, family, order, class, phylum, kingdom, and domain. For example, the fungi belong to the kingdom Fungi, the animals to the kingdom Animalia, and the plants to the kingdom Plantae.

Latin names at the species level are "binomials"—that is, they consist of two parts, the name of the genus (plural: genera) followed by the name of the species ("specific epithet"). Ultimately, these two parts indicate the placement of a particular organism in a hierarchical classification system. In technical publications, the binomial is followed by the name (or an abbreviation of the name) of the person (or persons) who first described the species (called the "authority"). For example, the complete taxonomic name of white oak is *Quercus alba* L., the "L." indicating that this species was first described by Carl Linnaeus (1707–1778), the Swedish botanist who also first developed this classification system, which is named after him. In the following table, some entries are listed by genus name only, followed by the designation "spp."; this denotes several similar species in that genus, which we refer to as a group in the text. Our reference for Latin names is the *Atlas of the Vascular Plants of Arkansas* by Gentry et al. (2013; see Reading List).

FOREST TREES

COMMON NAME	LATIN NAME	HABITAT
Ailanthus	*Ailanthus altissima* (Mill.) Swingle	Invasive Asian tree
Ash, white	*Fraxinus americana* L.	Upland oak–hickory forests
Basswood, American	*Tilia americana* L.	Mesic ravines
Beech	*Fagus grandifolia* Ehrh.	Mesic ravines
Birch, river	*Betula nigra* L.	Stream bottoms
Black gum	*Nyssa sylvatica* Marshall	Upland forests and ravines
Box elder	*Acer negundo* L.	Stream bottoms and disturbed forests
Butternut	*Juglans cinerea* L.	Uncommon in upland forests
Cedar, red	*Juniperus virginiana* L.	Rock outcrops and disturbed forests
Cherry, black	*Prunus serotina* Ehrh.	Second-growth forests
Cucumber tree	*Magnolia acuminata* (L.) L.	Mesic ravines
Cypress, bald	*Taxodium distichum* (L.) Rich.	Wetlands (south of the Ozarks proper)
Dogwood	*Cornus florida* L.	Subcanopy tree in upland forests
Elm, winged	*Ulmus alata* Michx.	Upland oak–hickory forest and disturbed areas
Fringe tree	*Chionanthus virginicus* L.	Stream bottoms and rock ledges
Hackberry	*Celtis occidentalis* L.	Stream bottoms and disturbed forests
Hackberry, dwarf	*Celtis tenuifolia* Nutt.	Rock outcrops and glades

COMMON NAME	LATIN NAME	HABITAT
Hemlock, Canadian	*Tsuga canadensis* (L.) Carrière	Eastern Appalachian forests (not found in the Ozarks)
Hickory, bitternut	*Carya cordiformis* (Wangenh.) K.Koch	Stream bottoms
Hickory, mockernut	*Carya alba* (L.) Nutt. ex Elliott	Upland oak–hickory forests
Hickory, pignut	*Carya glabra* (Mill.) Sweet	Upland oak–hickory forests
Hickory, shagbark	*Carya ovata* (Mill.) K.Koch	Upland oak–hickory forests and stream bottoms
Hickory, Texas	*Carya texana* Buckley	Dry, rocky uplands
Hornbeam	*Carpinus caroliniana* Walter	Subcanopy tree along head-water streams
Ironwood	*Ostrya virginiana* (Mill.) K.Koch	Subcanopy tree in upland forests
Lepidodendron	*Lepidodendron* spp.	Extinct tree of ancient Ozark forests
Locust, black	*Robinia pseudoacacia* L.	Disturbed areas and roadsides
Locust, honey	*Gleditsia triacanthos* L.	Disturbed areas and roadsides
Magnolia, umbrella	*Magnolia tripetala* (L.) L.	Small, multi-stemmed tree of headwater streams
Maple, red	*Acer rubrum* L.	Upland oak–hickory forests
Maple, sugar	*Acer saccharum* Marshall	Mesic ravines and fertile uplands
Morel	*Morchella* spp.	Edible fungus found in upland oak–hickory forests during spring
Oak, black	*Quercus velutina* Lam.	Upland oak–hickory forests

COMMON NAME	LATIN NAME	HABITAT
Oak, blackjack	*Quercus marilandica* Münchh.	Dry uplands and shale glades
Oak, bur	*Quercus macrocarpa* Michx.	Margins of prairies, mostly on the northern side of the Ozarks
Oak, chinquapin	*Quercus muehlenbergii* Engelm.	Upland oak–hickory forests on calcium-rich soils
Oak, northern red	*Quercus rubra* L.	Upland oak–hickory forests and some stream bottoms
Oak, pin	*Quercus palustris* Münchh.	Poorly drained areas
Oak, post	*Quercus stellata* Wangenh.	Upland oak–hickory forests
Oak, scarlet	*Quercus coccinea* Münchh.	Present on the eastern edge of the Ozarks
Oak, Shumard	*Quercus shumardii* Buckley	Upland oak–hickory forests and some stream bottoms
Oak, southern red	*Quercus falcata* Michx.	Upland oak–hickory forests and some stream bottoms
Oak, white	*Quercus alba* L.	Upland oak–hickory forests
Ozark chinquapin	*Castanea pumila* var. *ozarkensis* (Ashe) G.E. Tucker	Dry upland ridges and rock outcrops
Pawpaw	*Asimina triloba* (L.) Dunal	Thicket-forming subcanopy tree on rich soils
Pine, jack	*Pinus banksiana* Lamb.	Fire-prone northern forests (not present in the Ozarks today)
Pine, shortleaf	*Pinus echinata* Mill.	Dry upland forests on ridgetops
Plum, Mexican	*Prunus mexicana* S.Watson	Upland forests

COMMON NAME	LATIN NAME	HABITAT
Redbud	*Cercis canadensis* L.	Small flowering tree of rich woods and roadsides
Serviceberry	*Amelanchier arborea* (F.Michx.) Fernald	Subcanopy tree in upland forests
Spruce, white	*Picea glauca* (Moench) Voss	Boreal forest tree no longer present in the Ozarks
Sweetgum	*Liquidambar styraciflua* L.	Stream bottoms and disturbed areas
Sycamore	*Platanus occidentalis* L.	Stream bottoms and disturbed areas
Walnut, black	*Juglans nigra* L.	Stream bottoms and disturbed areas

SHRUBS AND VINES

COMMON NAME	LATIN NAME	HABITAT
Azalea	*Rhododendron prinophyllum* (Small) Millais	Flowering shrub on dry, exposed sites
Black haw	*Viburnum prunifolium* L.	Upland forests
Buckeye, red	*Aesculus pavia* L.	Stream bottoms and rich soils
Deerberry	*Vaccinium stamineum* L.	Low shrub in upland forests
Farkleberry	*Vaccinium arboreum* Marshall	Large shrub on rocky, exposed sites
Grape, summer	*Vitis aestivalis* Michx.	Vine in all Ozark habitats
Grape, winter	*Vitis cinerea* (Engelm. In A. Gray) Engelm. ex Millardet	Vine in all Ozark habitats

COMMON NAME	LATIN NAME	HABITAT
Greenbrier	*Smilax* spp.	Thorny vine in disturbed areas
Honeysuckle, Japanese	*Lonicera japonica* Thunb. ex Murray	Invasive Asian vine in disturbed habitats
Huckleberry	*Gaylussacia* spp.	Blueberry-like shrub in the Appalachian region
Huckleberry, low-bush	*Vaccinium pallidum* Aiton	Low, thicket-forming shrub in upland areas
Hydrangea	*Hydrangea arborescens* L.	Moist, boulder-strewn stream bottoms
Possum haw	*Ilex decidua* Walter	Stream bottoms and disturbed areas
Poison ivy	*Toxicodendron radicans* (L.) Kuntze	Ground cover and climbing vine in all habitats
Privet, Japanese	*Ligustrum japonicum* Thunb.	Invasive Asian shrub in disturbed habitats
Rattan vine	*Berchemia scandens* (Hill) K.Koch	Tree-climbing vine in upland forests
Spicebush	*Lindera benzoin* (L.) Blume	Common shrub in stream bottoms
Sumac, fragrant	*Rhus aromatica* Alton	Low shrub with a coffee fragrance, found on rocky sites
Sumac, smooth	*Rhus glabra* L.	Thicket-forming shrub in disturbed areas
Sumac, winged	*Rhus copallinum* L.	Thicket-forming shrub in disturbed areas
Virginia creeper	*Parthenocissus quinquefolia* (L.) Planch.	Vine in all Ozark habitats
Witch-hazel	*Hamamelis virginiana* L.	Multi-stemmed shrub in stream bottoms
Witch-hazel, vernal	*Hamamelis vernalis* Sarg.	Multi-stemmed shrub with fragrant flowers

WILDFLOWERS, HERBS, AND FERNS

COMMON NAME	LATIN NAME	HABITAT
Alum root	*Heuchera americana* L.	Moss-covered boulders
Baneberry	*Actaea pachypoda* Elliott	Moist upland soils
Bellwort	*Uvularia* spp.	Moist upland soils
Blazing star	*Liatris* spp.	Exposed rock outcrops and ledges
Bloodroot	*Sanguinaria canadensis* L.	Rich soils in upland forests
Cardinal flower	*Lobelia cardinalis* L.	Moist soils along streams
Cattail, common	*Typha latifolia* L.	Roadside ditches
Dutchman's breeches	*Dicentra cucullaria* (L.) Bernh.	Moist soils in stream bottoms
Fern, Christmas	*Polystichum acrostichoides* (Michx.) Schott	Common throughout upland forests
Fern, maidenhair	*Adiantum pedatum* L.	Moist, rich soils
Fern, polypody	*Polypodium virginianum* L.	Moss-covered boulders and tree trunks
Fern, walking	*Asplenium rhizophyllum* L.	Moss-covered boulders
Fire pink	*Silene virginica* L.	Moist soils of stream bottoms
Ginger, wild	*Asarum canadense* L.	Moist soils of stream bottoms
Ginseng	*Panax quinquefolius* L.	Rich soils in old-growth forests
Goosefoot	*Chenopodium berlandieri* Moq.	Crop cultivated by early Native Americans
Hepatica	*Anemone americana* (DC.) H.Hara	Rich soils in old-growth forests

COMMON NAME	LATIN NAME	HABITAT
Horsemint	*Blephilia ciliata* (L.) Benth.	Open upland forests
Indian pipe	*Monotropa uniflora* L.	Moist forest soils
Iris, crested	*Iris cristata* Sol. ex Aiton	Rich soils in upland forests
Jack-in-the-pulpit	*Arisaema triphyllum* (L.) Schott in Schott & Endl.	Rich, moist soils of mature forests
Jewelweed	*Impatiens capensis* Meerb.	Moist soils along streams
Joe-pye weed	*Eutrochium purpureum* (L.) E.E.Lamont	Roadside ditches
Ladies' tresses orchid	*Spiranthes* spp.	Wet seepage areas
Ladyslipper, Kentucky	*Cypripedium kentuckiense* C.F. Reed	Rare in rich soils of old-growth forests
Ladyslipper, showy	*Cypripedium reginae* Walter	Rare in rich soils of old-growth forests
Ladyslipper, yellow	*Cypripedium parviflorum* Salish.	Rare in rich soils of old-growth forests
Liverwort	Order Marchantiales	Wet rocks on stream banks
Lobelia, great blue	*Lobelia siphilitica* L.	Wet seepage areas
Mayapple	*Podophyllum peltatum* L.	Upland forests
Milkweed, four-leaved	*Asclepias quadrifolia* Jacq.	Open upland forests
Monkeyflower, sharp-wing	*Mimulus alatus* Alton	Wet seepage areas
Monkeyflower, yellow	*Mimulus floribundus* Douglas ex Lindl.	Rare in wet seepage areas

COMMON NAME	LATIN NAME	HABITAT
Moss, broom	*Dicranum* spp.	Moist rocks and stream banks
Moss, fern	*Thuidium delicatum* W.P. Schimper in B.S.G.	Moist rocks and stream banks
Moss, haircap	*Polytrichum* spp.	Moist rocks and stream banks
Moss, white cushion	*Leucobryum glaucum* (Hedw.) Angstr. in Fries	Rocks and exposed soil
Ox-eye daisy	*Leucanthemum vulgare* Lam.	Naturalized in roadsides and disturbed areas
Paintbrush, Indian	*Castilleja coccinea* (L.) Spreng.	Glades and ledges
Partridgeberry	*Mitchella repens* L.	Moist, moss-covered boulders
Penstemon, foxglove	*Penstemon digitalis* Nutt. ex Sims	Open forests and meadows
Phlox, blue	*Phlox divaricata* L.	Upland forests
Prickly pear	*Opuntia humifusa* (Raf.) Raf.	Glades and ledges
Puccoon	*Lithospermum canescens* (Michx.) Lehm.	Glades and ledges
Rattlesnake plantain, downy	*Goodyera pubescens* (Willd.) R.Br. in Alton & W.T.Alton	Upland forests
Shining clubmoss	*Huperzia lucidula* (Michx.) Trevis.	Rare in mesic forests
Solomon's seal	*Polygonatum biflorum* (Walter) Elliott	Moist upland forests
Spring beauty	*Claytonia virginica* L.	All forests and disturbed areas

COMMON NAME	LATIN NAME	HABITAT
Sunflower	*Helianthus annuus* L.	Crop cultivated by early Native Americans
Trillium, toad-shade	*Trillium sessile* L.	Throughout upland forests
Toothwort	*Cardamine concatenata* (Michx.) O.Schwarz.	Upland forests
Trout lily, white	*Erythronium albidum* Nutt.	Upland forests
Trout lily, yellow	*Erythronium rostratum* W.Wolf	Stream bottoms
Verbena	*Verbena urticifolia* L.	Glades and exposed ledges
Violet, birdfoot	*Viola pedata* L.	Upland forests
Violet, three-lobed	*Viola palmata* L.	Upland forests and stream bottoms
Wild petunia	*Ruellia humilis* Nutt.	Moist soil of stream bottoms

FUNGI, PATHOGENS, AND PARASITES

COMMON NAME	LATIN NAME	HABITAT
Agaric mushroom	Family Agaricaceae	Leaf litter
Beech scale	*Cryptococcus fagisuga* Lindinger	Beech bark (but not yet in the Ozarks)
Bird's nest fungi	Family Nidulariaceae	Rotten wood and stumps
Bolete mushroom	*Boletus* spp.	Leaf litter
Bracket fungi	*Polyporus* spp.	Stumps and logs
Chanterelle mushroom	*Cantharellus* spp.	Leaf litter
Chestnut blight	*Cryphonectria parasitica* (Murrill) Barr	Ozark chinquapin bark

COMMON NAME	LATIN NAME	HABITAT
Cedar-apple rust	*Gymnosporangium juniperi-virginianae* Schwein.	Cedar foliage and apple leaves
Chicken-of-the-woods	*Laetiporus sulphureus* (Bull.) Murrill	Stumps and rotten wood (especially oak)
Coral fungi	Family Clavariaceae	Rotten wood and leaf litter
Cup fungi	Family Pezizaceae	Rotten wood
Emerald ash borer	*Agrilus planipennis* Fairmaire	Ash cambium (recently arrived in the Ozarks)
Horsehair fungi	*Marasmius* spp.	Leaf litter
Leaf gall	Many different organisms	Living leaves
Leaf miner	Order Lepidoptera	Living leaves
Locust, thirteen- and seventeen-year	*Magicicada* spp.	Underground, emerging adults damage shoots
Mistletoe	*Phoradendron leucarpum* (Raf.) Reveal & M.C.Johnst.	Infesting branches of oaks and other trees
Phomopsis gall	*Phomopsis* spp.	Branches of hickory and sometimes oak
Puffball	*Lycoperdon* spp.	Leaf litter
Red oak borer	*Enaphalodes rufulus* (Haldeman)	Red oak bark and cambium
Tar spot	*Rhytisma* spp.	Living leaves
Thelephore fungi	Family Thelephoraceae	Rotten wood and stumps
Stinkhorn	Family Phallaceae	Leaf litter and soil
Water mold	*Phytophthora* spp.	Soil and roots of living trees

GLOSSARY

Advanced reproduction. Suppressed tree seedlings or saplings growing on the forest floor that can be released and grow to become part of the forest canopy if the existing forest is cut or subjected to extensive sorm damage or disease mortality.

Alluvial flat. The relatively level area along streams and rivers where frequent overflow deposits keep the landscape relatively flat and fertile in comparison to valley slopes on either side.

Altithermal. The period from nine thousand to six thousand years ago when, geologists believe, the North American climate was slightly warmer and drier than in recent times.

Angiosperms. Broad-leaved trees such as oaks and maples, and other vascular plants such as daisies and orchids, that have their seeds in a closed ovary.

Arcto-Tertiary flora. Our deciduous forest trees' ancient ancestors that inhabited the high-latitude regions of a single, combined continental landmass about fifty-five million years ago, when the Earth's climate was considerably warmer than it is now.

Asa Gray disjunction. The presence of similar species of trees and other plants, such as witch-hazel and chinquapin, in both eastern North America and East Asia, although they do not occur anywhere in between—a pattern first described by the botanist Asa Gray.

Ascomycetes. A major group (taxonomic class) of fungi that contains very few species that produce the conspicuous fruiting bodies we recognize as mushrooms (though the highly-sought-after morel is an exception).

Bankfull flow. The volume of flow in a stream that just comes to the top of the natural banks of the stream without spilling over onto the adjacent alluvial terrain.

Baseflow. The relatively low flow that continues in a stream after a long period without rain; this flow is derived entirely from water seeping out from where it is stored in the pore spaces of streamside sediments.

Basement rocks. The much older granite-like rocks that lie beneath the sedimentary rocks that now occur at the surface over much (but not all) of the Ozark region.

Basidiomycetes. A major group (taxonomic class) of fungi including most of the species that produce the conspicuous fruiting bodies we identify as mushrooms; the spores are borne on many small structures ("basidia") that give this group its name.

Bathymetry. Measurement of the subsurface topography of a body of water.

Bedload. The relatively coarse sediment load of a flowing river or stream that is transported downstream by being pushed along the bottom, with transport often episodic during high flow events.

Brachiopod. An ancient bivalve mollusk with a pair of shells that look like the Shell Oil trademark and that are commonly found as fossils in Ozark limestone formations.

Bract. Modified leaf associated with the base of a true flower or inflorescence and often mistaken for a flower petal; the showy white "flowers" of dogwood are bracts that are not actually part of the flower cluster they surround.

Branch node. The place where a branch intersects with the trunk of a tree; branch nodes on pine logs consist of branch junctions that are symmetrically arrayed around the circumference of the trunk and can be especially conspicuous on such logs.

Buck rub. Damage to the base of a tree sapling caused by male deer pushing on them in an effort to remove the velvet skin that covers maturing antlers.

Cambium. Cylindrical layer of living cells located outside the xylem and inside the phloem; these cells grow to account for the annual diameter increase of a tree and thus produce what we refer to as "tree rings."

Canadian Shield. The large area of exposed, very ancient basement rocks that forms the central core of the North American continent; similar basement rocks extend underneath the sedimentary layers of the Ozarks and are exposed at the surface in the St. Francois Mountains of southeastern Missouri.

Canebrake. Dense thicket composed of closely spaced stems of the bamboo-like native plant known as "big cane."

Carbon dating. Method that uses the ratio of radioactive C^{14} to common C^{12} to determine how long an organic deposit has been isolated from the generation of radioactive carbon in the atmosphere, where it is exposed to cosmic rays; the decay rate of radioactive carbon is such that the technique can be used only on samples less than fifty thousand years old.

Centrifugal force. The force required to make water and sediment veer from their direct downstream course; the interaction of centrifugal force and water pressure plays an important role in stream-bank erosion.

Chert. Hard, brittle mineral formed of glassy silica that often occurs as angular nodules within limestone formations.

Cleistogamous flowers. Late-season, self-pollinating flowers that never open, produced by plants such as violets.

Compound leaves. Leaves consisting of several separate leaflets arranged along the sides of an extension of the petiole in either a "pinnate" (as in hickories) or "palmate" (as in buckeyes) fashion.

Concretion. Hard spherical or egg-shaped mineralized body in sandstone or siltstone where minerals dissolved in groundwater flowing through the rock have precipitated around a nucleus that triggered the process at the time when the rock body was consolidated during its formation.

Conk. Relatively large and usually hard fungal fruiting body that appears as a semicircular shelf or "bracket" attached to the trunks of dead or dying trees.

Coppice. Clump of trunks originating from the stump of a tree that has been previously cut in logging or broken as a result of storm damage.

Core. Long, thin, cylindrical sample obtained by drilling into a tree or sedimentary deposit with a coring device; a sample column is retrieved when the drill is withdrawn.

Cork cambium. Layers of living tissue within the bark of a tree that generate new bark tissue.

Corm. A bulb-like underground stem characteristic of some wildflowers.

Cotyledon. The embryonic leaf or pair of leaves of a plant embryo that emerge and enlarge as a seed germinates; plants are divided into "dicots" (such as oaks and maples) that produce a pair and "monocots" (such as grasses and lilies) that produce one.

Crevice cave. Deep, cave-like opening generated when hard sedimentary rocks such as sandstone fracture along rectangular sets of joints and begin to slide downhill under the effects of gravity.

Crinoid. Sea animal that consists of a basal attachment, a long tubular stem, and a cluster of filter-feeding arms that extract food from the surrounding water; the intact creature looks like a flower, so it is known as a "sea lily."

Cross-bedding. Nested sets of concave upward laminations in sandstone deposits that represent the movement of sand grains down the back side of dune-like features; cross-bedding is characteristic of water flowing over deformable sand bodies.

Deadfall. The fallen trunk of a tree that has died and remained standing; as the roots slowly decay in contact with moist soil, the tree topples over from its own weight.

Diatoms. Microscopic algae that photosynthesize in the upper levels of seas; they are enclosed in cell walls that contain silica ("shells"), which are deposited on the ocean floor and form chert nodules in limestone or siltstone formations.

Diffuse-porous wood. The wood of trees such as pine and maple, in which growth rings are composed entirely of small-diameter, water-conducting, tubular cells.

Dioecious plant. Any plant that produces male and female flowers on separate individuals.

Dolomite. A rock type similar to limestone and likewise subject to dissolution by groundwater; some of the calcium atoms in the originally deposited limestone mineral crystals have been replaced by atoms of magnesium.

Edaphic conditions. Conditions related to the soil, such as moisture content, aeration, drainage, and fertility.

Elaiosome. A small body consisting of an oil-rich substance, attached to the seeds of some wildflowers, that causes ants to be attracted to the seeds—and then to help disperse them.

Epikarst. The convoluted, buried surface layer developed by groundwater dissolution of rock when limestone or dolomite lies underneath the soil layer.

Epiphyte. A plant, such as the polypody fern and various bryophytes, that grows in the branches of trees.

Evulsion. The process whereby a stream establishes a new channel by breaking out of its banks at a location and flowing parallel to its former course for some distance.

Exclusion stage. The stage in forest regeneration when a dense young forest composed of seedlings and released trees from advanced regeneration begins to suffer severe mortality as a result of competition that will determine the ultimate composition of the mature forest.

Fire scar. A triangular section of exposed wood bordered by callus ("scar") growth indicating where a relatively low-intensity fire has killed the bark on the (usually uphill) side of the trunk where a thicker layer of fuel had been allowed to accumulate.

Forest association. Broad region inhabited by recognizable assemblages of tree species (for example, oak and hickory) as defined by the U.S. Forest Service and various academic institutions.

Forest forensics. The use of physical investigative techniques, ranging from qualitative inspection of tree growth form to quantitative tree-ring series analysis, to the study of the history of a forested area.

Forest growth model. Numerical simulation of forest stand history by computing the rates of diameter and height growth determined by calibrated growth and mortality rates, as influenced by climate, competition, and availability of mineral resources.

Geomorphology. The scientific discipline that encompasses the study of natural forces that govern the development of landscapes.

Gills. Thin, blade-like structures that radiate from the central stalk on the underside of the cap of some mushrooms.

Glade. Habitat in an otherwise forested landscape where shallow soil, exposed bedrock, and/or extreme infertility result in open conditions with widely spaced shrubby trees or no trees at all.

Gymnosperms. Coniferous trees such as pine and juniper found in Arkansas (and other examples, such as cycads, that are not found in our area).

Hardpan. A shallow layer of impermeable clay or compact sediment that impedes drainage of soils.

Heartwood. The cylinder of wood within the center of a tree that has been infused with resins and no longer conducts sap upward into the crown of the tree.

Hyphae. The thread-like structures that form the network that makes up the main body of fungi.

Isotopic analysis. Quantitative analysis of the ratio of elements of different isotopic weights but similar chemical properties as they are influenced by metabolic processes in organisms; often, this involves the ratio of O^{18} to O^{16} or C^{13} to C^{12}, but it can be extended to many other elements such as strontium and sulfur.

Joint. A seam-like discontinuity in layers of brittle rock such as sandstone or limestone; joints usually occur in a rectangular pattern and are generated over long spans of geologic time in response to stress relief as erosion removes the weight of overlying rock.

Juvenile leaves. Unusually large and strangely shaped leaves of trees, found on the basal shoots and on vigorous seedlings, that can contrast with the familiar leaves of the species in question.

Karst. Landform developed where bedrock consists of limestone or dolomite that can be dissolved by rainwater, creating enlarged joints in rocks and sinkholes; important characteristics of karst landscapes include the presence of sinking streams and the capacity for rapid migration of any contaminants introduced into the subsurface.

Kopje. A South African term for piles of bedrock boulders that are exposed by erosion on African plains.

Lichenized fungi. The organism we commonly refer to as a "lichen" is a combination of a fungus and an alga living together to form a composite structure that is different from what we commonly recognize as either type of organism.

Lignin. Extremely tough organic polymer that forms the main part of wood and bark tissue in trees and other plants.

Macrofossils. Plant and animal remains large enough to be recognized and described with the naked eye, as opposed to "microfossils" such as pollen grains and diatoms.

Mast. Term for the fruit production by nut-bearing trees; the fluctuations in the size of the nut crop from one year to the next that often occur (known as "masting") are thought to be a way of discouraging seed predation.

Megafauna extinction. The rapid disappearance of a wide range of large mammals, such as horses and mammoths, in North America about thirteen thousand years ago.

Mesic. Adjective describing habitats in sheltered areas, such as the base of a bluff or the lower side of a ravine, where soil moisture is available throughout the year.

Mesification. The process in which lack of fire or other disturbance allows shade-tolerant tree species that are found in mesic habitats to invade drier sites normally inhabited by less shade-tolerant trees such as oaks.

Microhabitat. The local environmental conditions of a particular site that influence the growth of a tree or other plant; examples include small depressions, seepage areas, and local outcrops of exposed rock.

Midden. A mound-like deposit of organic debris collected by rodents that can provide an effective sample of everything growing in the local area at the time; middens frequently preserve these samples by virtue of being built in protected rock crevices and saturated with (preservative) rodent urine.

Milankovitch cycle. Rhythmic changes in the Earth's climate in cycles of twenty thousand, forty thousand, and a hundred thousand years, first identified (by Serbian mathematician Milutin Milanković) as driven by cyclic changes in the shape of the Earth's orbit around the Sun, and later proven to have caused Pleistocene ice ages.

Mineral soil. Soil beneath the leaf litter and humus that usually covers the forest floor; this layer contains some organic material but consists mostly of weathered clay and rock fragments.

Monoecious plant. A plant that produces male and female flowers on the same individual.

Mosaic. Adjective describing a forested environment that consists of patches of forest with different species and structure such that no single group of species dominates throughout.

Mycelium. The entire mass of hyphae that make up the body of a fungus; a fruiting body (such as a mushroom) is a highly organized portion of the mycelium.

Mycoheterotroph. A plant that does not photosynthesize on its own and lives off the resources of surrounding trees or other plants through connections provided by a network of fungal hyphae.

Mycorrhiza. A symbiotic association involving a fungus and the roots of a plant; many plants, ranging from orchids to forest trees, require these associations in order to effectively absorb nutrients from the soil. There are two types: "endomycorrhizae" penetrate the cell structure of the root, and "ectomycorrhizae" form a sheath around the outside of the root tip.

Myrmecochory. The seed-dispersal strategy used by some forest wildflowers whereby seeds are equipped with an oil-rich body (elaiosome) that prompts ants to act as seed-dispersal agents.

Nitrogen fixation. The process whereby bacteria associated with the roots of plants (such as those in the pea family) can convert the inert gaseous nitrogen from the atmosphere to a molecular form that can be readily absorbed as an essential nutrient by plants.

Overkill hypothesis. The theory that Native Americans arriving in North America for the first time encountered large animals such as mammoths and ground sloths, which had never been hunted before and were therefore easily exterminated.

Paleozoic era. A long period in Earth's history, extending from 542 million years ago (when larger life forms were just appearing) to 251 million years ago (just after the Coal Age), during which the sedimentary rocks now exposed at the surface in the Ozarks were deposited.

Pangaea. An assembly of several continents that were driven together by continental drift many hundreds of millions of years ago and began to separate into the individual continents we see today about three hundred million years ago; the sedimentary rocks now exposed in the Ozarks were deposited when North America was a part of this supercontinent.

Petiole. Technical term for the stem of a leaf.

Phloem. Vascular tissue in the trunk of a tree, located outside the water-conducting xylem and underneath the bark, that conducts nutrients downward to the roots from the tree canopy above.

Pit-and-pillow topography. Array of small hollows and low mounds characterizing the forest floor in old-growth forest; when large trees are overturned, soil is pulled up with the roots, eventually producing the pits and mounds as the wood decays away.

Plow berm. Low shelf of soil that develops on the downhill side of a plowed field as a result of soil slumping after the plow overturns the lowermost row of soil.

Plunge pool. Water-filled hollow at the base of a waterfall, excavated by erosion from impacts of falling water and sediment over time.

Point bar. Sand or gravel bar that develops on the inside of a stream or river meander.

Pollen profile. The vertical distribution of tree, shrub, and herb pollen identified to plant genus in a column of sediment recovered by drilling into lake or pond sediments (or obtained by sampling from an exposed vertical excavation wall).

Radiometric dating. Method that uses the known decay rates of long-lived radioisotopes such as uranium to estimate the age when geologic deposits were formed; dating is done as with C^{14} but can extend over the entire 4.5 billion years of Earth's history.

Release event. Process whereby a tree growing in the understory of an established forest experiences an increase in growth rate when an event such as logging or damage to an adjacent overstory tree provides improved growing conditions.

Relict. Adjective describing a tree or other plant species, found today in our area, that survives from a former time and a different climate in local habitats where it still finds suitable growing conditions.

Rhizome. Thickened, root-like stem that serves as an elongated structure capable of lateral underground growth, which allows colonies of some wildflowers to expand.

Ring-porous wood. The wood of trees such as oak and elm, in which the growth rings contain a layer of relatively large-diameter, water-conducting vessels that form continuously connected conduits.

Riparian forest. Forest that grows on the banks and floodplains of streams and rivers, often consisting of trees such as sycamore and sweetgum that are well adapted to this type of environment.

Root plate. The slab of roots and attached soil that occurs at the base of a tree when it is toppled by strong winds or exposed on an undercut stream bank.

Root sprouting. The habit of some trees and shrubs, such as beech and black gum, to generate new stems at some distance from the trunk by the growth of shoots originating from shallow roots.

Rosette (basal rosette). Growth form of some herbaceous plants in which leaves radiate outward from a stem near the ground surface.

Samara. Type of fruit produced by elms, maples, and other species that has flattened wings of fibrous tissue to enhance windborne dispersal.

Saprophyte. An organism that lives on and obtains its nourishment from dead organic matter, such as fallen leaves or rotten wood.

Sapwood. Layer of sap-conducting cells underneath the bark and cambium of a tree.

Savanna. Landscape composed of grassland and low shrubs with scattered individual trees.

Scape. Flower stalk or stem.

Shade leaves. Relatively thin and broad leaves of trees; these leaves occur on lower branches where they are not exposed to full sunshine.

Shale. Relatively soft and easily eroded sedimentary rock formed when mud settles to the bottom of a quiet body of water and is later consolidated into a sedimentary rock layer when buried under other sediments.

Sheetflow. Situation where falling rain flows downhill over the land surface in a continuous layer of water that is not concentrated into discrete channels.

Sinkhole. Surface depression formed when soil collapses into an underlying cavity created by the dissolution of limestone in karst terrains.

Spring ephemeral. Any forest wildflower species that produces flowers and completes most of its life cycle in the few weeks of early spring, before the trees overhead leaf out.

Stipe. The stalk or stem of a mushroom.

Stolon. Underground stem produced by some plants for lateral vegetative propagation; the common strawberry provides a familiar example.

Stomata. Small pores (singular: stoma) on the underside of leaves that allow the intake of carbon dioxide and the loss of water vapor as part of the photosynthetic process.

Subduction. The process in plate tectonics whereby the ocean floor descends under more buoyant continental crust, generating periodic earthquakes and volcanic eruptions, but the movement is otherwise so slow as to remain undetectable by all but the most precise laser-based measurements.

Subsidence. The process by which a local area of land surface sinks, either under the weight of sediments being deposited in that location or because of an increase in subsurface density produced by cooling of volcanic intrusion at lower depths.

Succession. The process in which species of trees, plants, and/or fungi change or succeed each other over time as conditions of their surroundings change. Most often this refers to the changing series of species that occurs as an abandoned field reverts to closed forest.

Sun leaves. The thick and waxy leaves that occur in the upper canopy of a tree when exposed to wind and full sunshine.

Supercooling. The process whereby raindrops falling through frigid air have a temperature well below freezing but remain liquid for lack of a particulate nucleus to trigger ice formation; such conditions are not common but can result in catastrophic ice-storm damage to forests when they occur.

Suppression. Situation in which an understory tree is forced into a period of extremely slow growth because of the dense shade cast by competing overstory trees.

Suspended sediment. Particles of fine sediment held in suspension in a river or stream by the natural turbulence of the flow.

Talus. Slope covered with rocky debris generated by the weathering of rocks located upslope.

Tannin. An organic substance contained in tree tissues that serves as a chemical defense against pathogens, which accounts for the bitter taste that makes some nuts less palatable to seed predators.

Thallus. The composite tissue formed by the combination of an alga and a fungus that we recognize as a lichen.

Thong tree. Tree that is bent over to the side on its lower trunk but then rises vertically. Such trees are said to be markers to point the way, created many years ago by pioneers and Native Americans, but are almost certainly the result of natural damage to the tree in its early life.

Tolerance. The relative ability of specific species of trees to grow under shaded conditions; trees are rated as "intolerant" (requiring full sunshine), "tolerant" (growing well in deep shade), or "mid-tolerant" (growing in partial shade).

Tree-ring studies. The analysis of the measured widths of growth rings as acquired from cores drilled out of trees or from cross sections of tree trunks; the series of ring widths can be used to infer the timing of past disturbances that affected growth or can be calibrated in terms of rainfall and stream discharge.

Vascular plants. All higher plants in which vascular tissue conducts water and nutrients through the plant structure (as opposed to mosses and lichens, which do not possess this type of tissue).

Venation. The pattern of veins within a leaf, which can be described as "palmate" (radiating outward like fingers from the leaf base) or "pinnate" (extending outward along the length of a central main vein).

Vernal pool. Shallow and biologically productive body of water that exists in spring under seasonally moist conditions but dries up during summer.

Witness tree. Trees identified by species and their locations with respect to a township boundary as recorded in General Land Office records and used to infer the composition of forests at the time of land settlement.

Woody lignotuber. The portion of the root system of shrubs that connects the aboveground stem to the root system and is capable of propagating underground; this part of the plant can generate new shoots that allow lateral expansion of the shrub colony.

Xylem. The upward water-conducting vessels and other cells found underneath the cambium and bark of a tree.

Zircon. Extremely hard silicate mineral that forms in volcanic eruptions and can withstand extreme cycles of erosion; the sedimentary origin of sandstones can be identified by the radiometric age of the zircons they contain, combined with knowledge of the places in North America where volcanic eruptions were occurring at that specific time.

READING LIST

We suggest the following books, all readily accessible, for readers who want to learn more about the various aspects of forest ecology that we have presented along the way. Although this is not a comprehensive bibliography for the subjects covered in this book, each of these references is a starting point for going deeper into the woods.

Robert A. Askins. 2014. *Saving the World's Deciduous Forests: Ecological Perspectives from East Asia, North America, and Europe*. Yale University Press, New Haven, Connecticut.

A comprehensive overview of the effort to preserve deciduous forests and the way in which various cultural heritages influence the conservation concept embodied in that effort.

E. Lucy Braun. 1950. *The Deciduous Forests of Eastern North America*. Facsimile edition, 1972, Hafner, New York City.

The fundamental overview of our deciduous forests, with classic photos of old-growth forest in the Ozarks and elsewhere. Braun also presents some of the earliest ideas about the history of our forests.

David R. Foster and John D. Aber. 2006. *Forests in Time: The Environmental Consequences of 1,000 Years of Change in New England*. Yale University Press, New Haven, Connecticut.

An overview of the changes that have influenced deciduous forests with many similarities to our own oak–hickory woodlands. Also illustrates how forest forensic techniques were applied to that effort.

Johnnie L. Gentry, George P. Johnson, Brent T. Baker, C. Theo Witsell, and Jennifer D. Ogle (Editors). 2013. *Atlas of the Vascular Plants of Arkansas*. University of Arkansas Herbarium, Fayetteville.

Distribution maps for all 2,892 native and naturalized vascular plants reported in Arkansas.

David George Haskell. 2012. *The Forest Unseen: A Year's Watch in Nature.* Penguin, New York City.

An investigation of the activities that occur in deciduous forests during the annual cycle of the seasons. Not exactly forensics as such, but a lot of information about what's going on in the background and underfoot in the forest.

Paul W. Nelson. 2010. *The Terrestrial Natural Communities of Missouri,* third edition. Missouri Natural Areas Commission, Jefferson City.

A detailed discussion of the conditions that influence natural habitats in the Missouri Ozarks.

Henry Rowe Schoolcraft. 1996. *Rude Pursuits and Rugged Peaks: Schoolcraft's Ozark Journal 1818–1819*, edited by Milton D. Rafferty. University of Arkansas Press, Fayetteville.

Descriptions of the Ozark forests and landscape by one of the earliest travelers in our region.

Kenneth L. Smith. 1986. *Sawmill: The Story of Cutting the Last Great Virgin Forest East of the Rockies.* University of Arkansas Press, Fayetteville.

Mostly a history of the timber industry in Arkansas, with some nice descriptions of the old-growth forests that were being cut.

Steven L. Stephenson. 2010. *The Kingdom Fungi: The Biology of Mushrooms, Molds, and Lichens.* Timber Press, Portland, Oregon.

A comprehensive review of fungi, slime molds, and lichens that is accessible to readers interested in learning more about these organisms.

Steven L. Stephenson. 2013. *A Natural History of the Central Appalachians.* West Virginia University Press, Morgantown.

Description of the deciduous forests of the Appalachians, with many concepts and discussions that apply to the Ozark and Ouachita regions as well.

Henry David Thoreau. 2009. *The Journal 1837–1861*, edited by Damion Searls. New York Review Books, New York City.

An annotated presentation of Thoreau's famous journal in which he describes his rambling investigations of the natural world in Massachusetts.

May Theilgaard Watts. 1975. *Reading the Landscape of America*, second edition. Nature Study Guild, Rochester, New York.

A now classic discussion of how various forensic techniques can be applied to natural-history investigations across North America.

Tom Wessels. 2005. *Reading the Forested Landscape: A Natural History of New England*. Countryman Press, Woodstock, Vermont.

An overview of how the history of forested plots in New England can be inferred from their appearance, illustrated with an artist's representation of various New England scenes.

Tom Wessels. 2010. *Forest Forensics: A Field Guide to Reading the Forested Landscape*. Countryman Press, Woodstock, Vermont.

A follow-up manual of forest forensic techniques with numerous photographs.

INDEX

Figures that appear outside the page span for the subject are indicated with an "*f*" after the page number.

Tables are indicated with a "*t*" after the page number.

Glossary entries are indicated in **bold**.

FRED PAILLET is an adjunct professor at the University of Arkansas, where he conducts research and supervises student projects related to geophysics, hydrology and paleoecology. He obtained his PhD in geophysical fluid mechanics at the University of Rochester in New York in 1974 and began research in the application of mathematical models to environmental processes as Assistant Professor of Geology at Wright State University in Ohio. In 1978 he joined the U.S. Geological Survey's National Research Program in Denver, Colorado, conducting environmental research projects at numerous sites in North America. His primary research focused on water infiltration in fractured crystalline rocks at field sites such as the Hubbard Brook Experimental Forest in New Hampshire and the Miner Experimental Forest in New York, and in karst rocks in the Florida Everglades.

While studying hydrologic processes at various sites, Paillet also considered the climate tolerances of trees in relating pollen and tree-ring data to historic and prehistoric environments. Because American chestnut was a principal tree in the historic forests of the northeast now nearly wiped out by an introduced blight, he made a major contribution in reconstructing the ecology of that supposedly lost and academically ignored species. Since then he has been closely involved with The American Chestnut Foundation and its monumental effort to breed blight-resistant American chestnut for reestablishment of that iconic species in its former Appalachian range.

After retiring from the U.S. Geological Survey, Fred held temporary appointments at the University of Maine and the University of Rennes, France, before moving to Arkansas in 2009. He has participated in long-term environmental studies at the Hubbard Brook Experimental Forest, resulting in several books along with numerous journal articles and other technical publications. His personal involvement in chestnut ecology and reconstruction of past climates has incidentally resulted in expeditions to the chestnut forests of southern Russia and the mountains of central China.

Paillet's Ozark connection began when the U.S. Forest Service contacted him with a request to investigate the ecology of our own native chestnut tree species, the Ozark chinquapin. His studies showed that Ozark chinquapin ecology bore many parallels with that of chestnut, while demonstrating that our chinquapin was a real tree distinctly different from the shrubby Allegheny chinquapin of the southern Appalachians. This experience helped generate an affection for the Ozark region in general that persisted long after that project ran its course.

Paillet is an avid participant in Ozark Society activities and is frequently encountered on Ozark trails with his sketchbook in hand.

www.ingramcontent.com/pod-product-compliance
Lightning Source LLC
Chambersburg PA
CBHW032326210326
41518CB00041B/1099

STEVE STEPHENSON is a research professor in the Department of Biological Sciences at the University of Arkansas in Fayetteville, where he teaches such courses as plant biology, forest ecology and plant ecology. Prior to moving to the University of Arkansas in 2003, he taught biology and ecology at Fairmont State University in West Virginia for 27 years. Stephenson earned both his MS and PhD from Virginia Polytechnic Institute and State University (Virginia Tech) in Blacksburg, Virginia. The research he carried out for both of his degrees involved studying the forest communities of the Central Appalachian Mountains. One aspect of his research was directed towards determining what trees replaced the American chestnut in those forests in which that species was once dominant. His research program on forest ecology continued during the years he spent at Fairmont State University and was relocated to the Ozark Mountains when he moved to the University of Arkansas.

While Stephenson was a graduate student at Virginia Tech, he was introduced to mycology (the study of mushrooms and other fungi) by Dr. Orson K. Miller. As a result, he developed a keen interest in both fungi and a group of fungus-like organisms called myxomycetes, also known as slime molds. Over the past 40 years, he has collected and studied fungi and slime molds on all seven continents and in every major type of terrestrial ecosystem. During the period of 2003 to 2009, Stephenson served as director of a project funded by the National Science Foundation to document the worldwide distribution of myxomycetes and other similar organisms. His graduate students at the University of Arkansas have earned their degrees carrying out research on subjects ranging from the ecological effects of introduced *Eucalyptus* in Kenya to the slime molds and related organisms associated with the tropical forests of Brazil. Several other graduate students have studied various aspects of Ozark forests, including the effects of prescribed burning on wood-decay and litter-decomposing fungi, the ecology of chinquapin and the plants associated with cedar glades.

Stephenson is the author or coauthor of 11 books and more than 400 book chapters and papers in peer-reviewed journals. His books include *The Kingdom Fungi: The Biology of Mushrooms, Molds, and Lichens* (Timber Press, 2010), *A Natural History of the Central Appalachians* (West Virginia University Press, 2013) and *Mushrooms of the Southeastern United States* (Timber Press, 2018). He was a Fulbright Scholar at Himachal Pradesh University in India in 1987 and was a William Evans Visiting Fellow at the University of Otago, Dunedin, New Zealand, in 2002. He was selected as a Fellow of the Mycological Society of America in 2010 and has received two short-term Fulbright Specialist Awards: one to Vietnam in 2014 and the other to India in 2017.